The Changing Atmosphere

A Global Challenge

JOHN FIROR

YALE UNIVERSITY PRESS
NEW HAVEN & LONDON

Published with assistance from the foundation established in memory of Philip Hamilton McMillan of the Class of 1894, Yale College

Designed by James J. Johnson and set in Palatino Roman type by The Composing Room of Michigan, Inc., Grand Rapids, Michigan.
Printed in the United States of America by Vail-Ballou Press, Binghamton, New York.

The paper in this book meets the guidelines for permanence and durability of the Committee on Production Guidelines for Book Longevity of the Council on Library Resources.

10 9 8 7 6 5 4 3 2 1

Library of Congress Cataloging-in-Publication Data

Firor, John.
The changing atmosphere : a global challenge / John Firor.
p. cm.
Includes bibliographical references.
ISBN 0-300-03381-8 (alk.

1. Atmosphere. 2. Acid rain—Environmental aspects. 3. Ozone layer depletion 4. Global warming. I. Title.

QC861.2.F57 1990 90-34518
363.73'92—dc20 CIP

Contents

Preface

This volume arose from a publisher's request that I expand into a book an essay I had written about atmospheric problems entitled "Interconnections and the 'endangered species' of the atmosphere" (in *Journal '84* [Washington, D.C.: World Resources Institute, 1984]). In the essay I had discussed the "big three" atmospheric issues—acid rain, ozone depletion, and climate heating—and had emphasized how the three problems were related.

Attempting to expand this theme into a book, I took advantage of invitations to speak about the atmosphere by using each chapter outline as the basis for a lecture. This approach allowed me to test my material both by observing the audience's reactions and by trying to answer questions that arose. Always a few people in any group are truly interested in the details of a technical subject, and they are a joy to address; explaining complex interactions and reasoning using the language of the educated nonscientist is a challenging and rewarding enterprise. But the question I received most often was not about detail; it was a request for a judgment. "All that science is okay," people seemed to say, "but tell us straight—are we in trouble or not?" I began to suspect that I should not simply add more technical details to the discussions of the essay: the book should address this commonly asked, ultimate question.

My suspicion was reinforced as I reviewed the current scientific writings about atmospheric changes. Several excellent and lengthy discussions of each of these atmospheric issues already exist, a few written by individual authors but most produced by distinguished international committees called together to summarize and evaluate the current state of knowledge. Such committees, while displaying expert judgment about the state of scientific knowledge, frequently fail to answer the question "Do we or don't we have a problem?" Scientists quite rightly seek strong foundations for their conclusions and avoid going beyond the firmly demonstrated and repeatedly verified. In addition, scientists have been, and have seen their colleagues, proved wrong often enough that they bend over backward to give the views held by even a small fraction of the community a respectable place in the review reports.

I see no reason to criticize scientists' caution or skepticism. Scientific progress can be uncertain and may become clear only years later, when the advance can be seen in a larger context. Surprises are common, and the credibility of the scientific community is always at stake. So in the long term, erring on the side of caution is appropriate. Indeed, these skeptical, careful habits have helped maintain the healthy scientific enterprise so valuable in generating knowledge about the world around us, and reports by select committees are likely to continue to be careful, conservative, and cautious.

But this custom also means that to the public or the policymaker, scientists' reports seem to emphasize mainly what is not yet known and what needs further study. This emphasis holds even if the committee agrees that a significant atmospheric change is under way. Hence the puzzlement of the public and the questions I received.

Even though the customs of academies and committees are quite familiar to me, I realized, while rereading older reports and catching up on newer ones, that many of the authors I know well hold much stronger, more definite views than emerge from their committee endeavors. Clearly, some feature of their experience and studies, some application of more general notions about the workings of the world, allows them to hold well-developed opinions about the severity of the problems in the air while simultaneously joining in with other scientists to produce reports that are seen by many as carefully noncommittal. Sometimes we see these scientists on television or quoted in the press and there learn that they firmly believe that we do indeed have a serious problem arising from our careless treatment of the atmosphere. But these glimpses are generally so brief that the public is hard-pressed to figure out why these scientists hold such views, or why there is a gap between official reports and unofficial worries.

It is my hope that this book can fill that gap. My aim has not been to produce just one more careful discussion of the state of the science surrounding acid rain, ozone depletion, and climate heating, but to show why so many workers in the field have become convinced that emissions into the atmosphere have indeed reached a serious level.

This book consists of seven chapters. The first introduces two overlooked facets of the atmosphere that are essential to understanding today's worries. The next two cover two of the three well-known atmospheric problems: acid rain and depletion of the ozone layer by chlorofluorocarbon production. Climate heating produced as infrared-trapping gases accumulate in the air is so firmly at the center of all these issues that chapters four and five are devoted to it. The final two chapters discuss what these problems mean to people, to decision-makers, and to civilization, and what can be done about them.

Each of the three "famous" atmospheric problems sounds different from the others: acid rain in Canada and the northeastern United States, ozone loss over the South Pole and elsewhere, and climate change over the globe. Although in most discussions of these problems solutions are proposed as though each was separate from the others, they must be considered in unison. Their sources are strongly interrelated, and all three are tightly connected to such national and global issues as the energy supply, Third World development, foreign trade, North-South equity, and the population explosion. By isolating these atmospheric problems, we miss fundamental aspects of the issues we face and the courses of action open to us. Especially, the opportunity to take steps that tend to solve atmospheric and other societal problems at the same time is a basic feature of the current situation.

I have also focused attention on a human time scale—a few decades or up to a century, with a few necessary references to earlier events. We are beginning to understand things that take place over hundreds of millennia or millions of centuries, such as the origin of ice ages, the impact of continental drift on climate, and the extinction of dinosaurs. All of these topics make exciting reading, and each bears in some way on today's atmospheric problems. But combining the slow, large-scale events with the theme of this book—the impact of people on the future of the atmosphere—would be misleading. A careful and perhaps entertaining look at a climate markedly different from our own during some distant era may tempt us to think, fatalistically, that whatever happens in the next few decades happens because we are at the mercy of cosmic forces beyond our control. In fact, we are at the mercy of our own actions, our own goals, our own organizations and governments, our own understanding of the impact of human activities, and our own energy and willingness to do something about them.

Finally, any review of the scientific basis for concern about the atmosphere, and especially if that review concludes, as I do, that the problems in the air are more serious than they may seem from daily observations of the world around us, must face the issue of whether we can do something about them. One of the reasons for concluding that the problems are more serious than they seem is that their roots go so deeply into the nature and history of human civilization, and this reason guarantees that no easy solutions exist. So I have described the solution in terms of a choice of future paths, in one of which we stabilize the composition of the atmosphere, rather than letting it continue to be modified accidentally by human activities. It seems obvious, even before one examines any detailed scientific predictions, that if the air continues to change its composition indefinitely, at a rate measured by a few human generations, we will eventually produce major changes in everything we do and in all other living things. Thus the alternative to stabilization of the air is a path in which we continuously adapt to the rapid changes we are forcing on the atmosphere and, in the process, devise methods to manage all natural systems.

ACKNOWLEDGMENTS

I have the privilege of serving on the staff of the National Center for Atmospheric Research, and the constant interaction with other NCAR scientists and with visitors has been the most important influence on my scientific understanding and appreciation of the complex system that is our atmosphere. This research organization not only provides an environment conducive to thinking and writing but also encourages its staff to take a broad view of atmospheric problems. NCAR has grown in the last quarter-century and developed its high place in the country's scientific establishment through the steady and intelligent sponsorship of the National Science Foundation. NCAR is operated by a group of North American universities—the University Corporation for Atmospheric Research, or UCAR—and this relationship has provided me with further opportunities to broaden my view of the nature of the atmosphere and the challenges inherent in attempting to understand the interaction of the atmosphere with society. All of these organizations have been helpful to me, but of course the opinions in this book, and any mistakes, should be attributed to me and not to NCAR, UCAR, or the National Science Foundation, or any of the kind individuals who read and commented on sections or the entire manuscript.

I would like especially to acknowledge my debt to Walter Orr Roberts, William W. Kellogg, and Stephen H. Schneider, three modern pioneers in the study and understanding of how the atmosphere responds to human activities, and to Michael H. Glantz, who studies how human activities respond to the atmosphere and its changes. They have taught me much, and the example they set of how to think clearly and work effectively on these scientific areas, despite great public, press, and political attention, is a valuable one.

Service on the Boards of Trustees of the Environmental Defense Fund and the World Resources Institute has not only brought me in contact with a number of talented people, it has also taught me that the task of converting advancing scientific knowledge into appropriate public policy and action may be even more difficult than making the scientific advances in the first place.

I also wish to acknowledge the help of the following people, who made specific and essential contributions: William Chandler, Ralph Cicerone, Kathleen Courrier, Judith Jacobsen, Andrew Scott, and Anne Firor Scott each read some or all of the chapters, sometimes including several successive drafts, and made critical comments and helpful suggestions. Justin Kitsutaka and the members of the NCAR graphics group prepared the figures.

The Changing Atmosphere

1. The Atmosphere and People

As children we learn slowly about the atmosphere. We feel it on our faces, we realize that we are breathing something. Later we fly kites, blow out candles, and see the results of destructive winds. These early observations lead us to think of the air as something a bit like a glass of water—a substance whose characteristics and composition are given to us by nature, something apart from us and unchanging. Gradually we add to this first impression the realization that we as individuals are also completely dependent on the atmosphere and must take care not to be very long without a fresh lungful.

These early deductions miss one overriding fact. We miss it, not because we are unobservant, but because we have no way of noticing changes that happen slowly over very long periods of time. The air, in fact, is not "something apart"; it is the creation of ourselves and all other living things. Each day large quantities of air are absorbed by living matter and processed in various ways. Each day large quantities of gases are released into the air by living material. Over geological time, the evolution of life has been intertwined with the evolution of the atmosphere; the characteristics and composition of our air depend on the pattern of life forms on earth and are completely different from what they would have been had life not been present, and indeed completely different from the atmospheres of our neighboring, lifeless planets, Mars and Venus.[1]

The earth coagulated out of the gas and dust drifting around the sun over four billion years ago, and life emerged a half-billion years later. Oxygen, which is a major and important constituent of the atmosphere, arrived rather late in the day, long after single-cell organisms appeared. The sequence of events seems to have been as follows: The first living matter used, as food, substances produced in the geological processes. For example, a small amount of hydrogen and larger amounts of hydrogen sulfide came from volcanoes, and these are substances that can supply energy to a cell. After a very long time, a new process, photosynthesis, evolved that used an abundant component of the air—carbon dioxide—and an abundant source of energy—sunlight—to create carbohydrates from carbon dioxide and water. As we all learn in high school science class, the waste product of this process is oxygen. Then over the slow history of the earth, oxygen accumulated in the air, and creatures that require oxygen—animals—evolved.[2] The full range of life as we see it now had then emerged. This process changed the distribution of many components of the atmosphere. Nitrogen reacts fairly slowly with other substances, but it is an essential part of living material, so the balance between nitrogen in the air, nitrogen in the ocean, and nitrogen in the solid earth is strongly modified by living things. Carbon dioxide cycles through the biosphere rapidly; in the process some is stored as limestone and marble, as gas and oil and coal, and as persistent organic matter in soils.

This series of events not only determined the composition of the air, it also modified the temperature of the earth. Carbon dioxide is an infrared-trapping gas—it absorbs radiation that would otherwise escape from the earth—and thus keeps the planet warm. Oxygen is not effective in trapping outgoing radiation. (In general, a symmetrical molecule such as oxygen, which consists of two identical atoms, has many fewer ways to rotate and vibrate, and thus absorb and emit radiation, than do molecules with complicated shapes such as carbon dioxide and water vapor.) So adding oxygen and subtracting carbon dioxide, as photosynthesis did, allowed the earth to be cooler than it would have been otherwise. Some carbon dioxide remains in the earth's atmosphere, regenerated by the respiration of plants and animals and supplied by emissions from volcanoes and hot springs. Water, evaporated from seas and lakes, also traps infrared radiation, so some degree of infrared trapping and climate warming continues to this day.

The relationship between the composition of the atmosphere and living matter is a close one. Nitrogen is not very reactive—the average nitrogen molecule spends ten million years in the air before being taken up by a biological process—but carbon dioxide makes the rounds about every six years. Even processes that appear at first glance to be chemical and mechanical frequently have a biological component. For example, carbon dioxide that disappears from the atmosphere for long periods of time by being incorporated in layers of rocks does not do so by purely chemical steps. It is first dissolved in sea water, then taken up by small creatures and made into shells, and the shells fall to the bottom and become part of the sediment. Both carbon and nitrogen are stored in coal deposits when living material is buried, and all sorts of organisms on the rock surface hasten the transformation of rocks into soil. Perhaps the most graphic illustration of the creation of the atmosphere by living things is the case of oxygen. This gas has been in the atmosphere in approximately its present concentration for more than a billion years, but were the production of oxygen by living things suddenly halted, the weathering of rocks would remove this oxygen from the air in a small fraction of that time, roughly four million years. Living things have regenerated the oxygen we need in the air hundreds of times in the history of the earth.

The air, therefore, far from being "something apart," is simply one component of a coupled, interacting system. All living things combine to establish the composition of the atmosphere; changes in the atmosphere affect life everywhere. In this book I discuss some aspects of this coupling, but I take the dominant theme one step further. To an increasing degree, people are no longer just one out of the millions of species interacting with the atmosphere. There are now so many of us, and each of us on the average wields so much power, that our influence on the whole global system is easily measured. When traveling by air over oceans or deserts and looking down on vast areas that are either uninhabited or sparsely inhabited, we may find it difficult to believe that people can change the earth. But in the atmosphere society has found, quite accidentally, what a Wall Street takeover specialist would call a "highly leveraged situation." The temperature of the earth's surface is determined by the composition of the air, but the parts of the air that make the greatest difference are not the major components, but rather a few "trace" constituents. Carbon dioxide, the gas most responsible for keeping the earth warm, occurs in the air in a concentration of less than 0.04 percent. To make a change in the warmth of the earth, therefore, we need not meddle with the entire five quadrillion tons of the atmosphere, but only with a much smaller fraction. The chemistry of the air that leads to acid rain or urban air pollution involves substances that appear in concentrations of a few parts per billion. And we measure the concentrations of ozone-destroying substances in parts per trillion.

Any discussion of the atmosphere today must treat this new and growing force that acts on the air: it must consider the human characteristics that cause us to rush into unknown and threatening changes, and it must search for realistic courses of action that will avoid pushing us into a far less than desirable future.

2. Acid Rain

Some years ago *National Geographic* published a new map of the United States. The map was actually many photographs, taken from earth-orbiting satellites, carefully pieced together into one grand picture, as if one were looking straight down on the country.[3] Rivers and lakes, mountains and deserts, forests and fields are easily distinguished, and most people seeing this picture for the first time admire the grander aspects of the landscape for a few moments and then marvel at the features they can recognize in the region in which they live.

On that photograph, in the southeastern part of the country, appears a small circular area, orange in color, that does not resemble any common feature of the landscape. Comparing the photo to a conventional map shows that the spot is not in or near any large city or anomalous geological feature such as a volcano or a granite dome. Indeed, one suspects that it is a defect in the film or a blemish from the production process.

But a visit to the area shows that the curious spot is real. Although the surrounding region—the southern Appalachian foothills of north Georgia and southeast Tennessee—consists of green farms and dense forests, the spot itself represents hundreds of square miles of eroded hills, a few stunted trees, and sparse vegetation. Two towns and an artificial lake are in the area and a four-lane highway passes through. People can be seen going shopping or to work. Activities seem normal, except they are carried out against a background more like a moonscape than a moist southern mountain terrain.

A roadside marker tells the traveler that copper ore was discovered here early in the nineteenth century and that, soon after, a smelter was built. Signs in the local museum tell the history of the smelter: how it was first operated, how it was closed for several years during the War Between the States, and how the operation of the mine, the smelter, and the coming of the railroad caused various population shifts. And the museum signs also tell of the sulfurous smoke from the smelter's early days that damaged trees, prevented logged-over areas from reseeding, and finally denuded the valley of vegetation.

The history of Copper Hill, Tennessee, holds several lessons for anyone studying acid rain today.[4] The first such lesson is that we do not necessarily learn from the past. English scientists as early as 1661 noted the influence of industrial emissions on the health of plants and people.[5] They suggested placing industries outside of towns and using higher smokestacks to spread the smoke into distant parts. (Even earlier, Queen Elizabeth had issued a proclamation forbidding the burning of coal in London while Parliament was in session.) Seventy-five years later a Swedish scientist noted that the "poisonous, pungent sulfur smoke" of a smelter "corroded the earth so that no herbs can grow around it." Later, in 1872, Angus Smith published a book in England called *Air and Rain: The Beginnings of a Chemical Climatology,* in which he described twenty years of field observation and research on the problem to which he gave the name "acid rain." He discussed many of the ideas we grapple with in studying today's acid-rain problem, and he described how sulfuric acid in the air corrodes metals and bleaches fabrics. Smith also put forward procedures for collecting and analyzing rain samples. Thirty years later, other English scientists showed that acidified rain inhibited plant growth and seed germination, as well as nitrogen fixation in soils.

Other incidents, anecdotes, and scientific studies from many countries covering hundreds of years tell the same story; society has long known that sulfur compounds in the air are agents of destruction. In literature, the notion that hell has a sulfurous smell goes back at least as far as Dante and probably to even earlier times. So the early miners and smelter operators at Copper Hill could have foreseen the permanent damage their activities would produce, but either they did not or they chose to ignore the possibility.

The second lesson also relates to the use of information. A visitor to this biologically burnt-out region receives not only a recounting of the area's history but also various interpretations of the past events. Some local historians, for instance, hold that the sulfur had little to do with today's desolation, that trees were cut to supply fuel for the smelter, and that this deforestation left the hillsides permanently bare. This explanation does not account for the thousands of hillsides in the eastern United States that have been extensively logged, some repeatedly, without becoming orange spots easily photographed from space. This interpretation teaches us that many people are unwilling to recognize the adverse effects of substances they have become accustomed to living with and on which, they are told, their jobs may depend. In this regard, many east Tennesseans resemble the inhabitants of Los Angeles, who for years denied that auto exhaust had anything to do with smog.

The final lesson to be learned from Copper Hill is quite different. The current owners of the mine and smelter have instituted a program to find tree and plant species that will, along with soil-treatment procedures, allow them to revegetate, even reforest, the area. The program is moving ahead, but the announcement of it generated opposition from a few local groups, who pointed out that they had for generations lived among and grown to love the barren red hills and would not welcome a radical change in scenery. Some refer to the area as a "beloved scar" and ask that it be left unchanged. The third lesson is that, given time, people can adapt to changing conditions, even convincing themselves that they are better off after the change, no matter what the rest of the world may think.

While evidence from Copper Hill and smelters around the world was firming up the case that airborne sulfur compounds could kill plants, other events were implicating sulfur in damage to people. At Copper Hill, the direct fumigation of the valley by sulfur dioxide seemed the likely source of most of the damage to vegetation. But as smokestacks there and elsewhere were built higher in hopes of cleaning up the air in the vicinity of the emission, the sulfur dioxide lingered longer in the atmosphere before returning to earth. In the process, some of it was oxidized into sulfates and sulfuric acid, both of which harm people when inhaled. In widely dispersed but dramatic episodes, people died by the tens or hundreds or thousands in the Meuse Valley, Belgium, in 1930; in Donora, Pennsylvania, in 1948; in London in 1952, 1953, and 1962; in New York City in 1953; and in a wide area of the eastern United States in 1966 during stagnant atmospheric spells, when sulfate accumulations in the air reached very high levels.[6] Less dramatically, nearly a century after Angus Smith noted acid rain's ill effects, it began to be noticed again that stone, concrete, paint, and steel deteriorated more rapidly in sulfurous cities.

Since some of these deadly episodes and examples of damage to buildings occurred at places far from sulfurous smelters, another, larger source of sulfur in the atmosphere had to be blamed. Such a source—fossil fuels—was not hard to find. Fossil fuels frequently contain small amounts of sulfur as an impurity; burning the fuel also burns the sulfur and produces gaseous sulfur dioxide. So much fossil fuel is used each day that, even with sulfur contents of only a percent or so, fuel combustion can place millions of tons of sulfur into the air each year. In industrial regions of Europe and the United States, sulfur concentrations in the air are now ten or twenty times those of the era before coal was burned.

Illness and haze in urban areas brought about attempts to reduce sulfur dioxide concentrations in or near large cities. Initially, the higher smokestacks were partially successful in sending the sulfur dioxide away from the city. Later, steps were taken to reduce, rather than just disperse, the harmful emissions. In some cases, fuel oil with a low sulfur content was substituted for high-sulfur coal. Devices were added to the smokestacks of new power plants to remove some of the sulfur from the smoke. These devices and fuel substitution brought emissions below what they would have been if no corrective actions had taken place, and sometimes the emissions dropped dramatically. In the United States, for example, where sulfur dioxide emissions had grown rapidly with the economy, corrective steps brought emissions down from a peak value of nearly 30 million metric tons a year in 1970 to below 25 million tons after 1980. Still, this rate is much higher than in preindustrial times and large enough to produce serious damage.[7] Globally, sulfur dioxide emissions are about 100 million tons annually and are probably still increasing as Third World countries hasten to take their place among industrial nations.

Given the centuries-old appraisal that sulfur in the air is damaging, and the occasional dramatic demonstration of this fact, it may seem strange that acid rain would emerge in our time as something of a surprise and that it would be so hard for people, industries, and countries to fix upon a remedy. But we need only remember Copper Hill to see how strongly people resist believing that a familiar condition—especially one tied to jobs and income—has negative long-term consequences. Then, too, the nature of the problem has changed. Most damage from smelters occurs within sight of the smokestack. In contrast, today's acid-rain problem is much more subtle: the sulfur compounds are more dilute, and even though widespread damage is suspected in lakes and forests, the source of the sulfur and the suspected damage are frequently far apart, and the link between them is complex and difficult to prove.

Volcanoes, swamps, and the sea all emit sulfur compounds into the air, and like the sulfur from smokestacks, these compounds can be transformed into sulfuric acid and sulfates. This fact raises a question: If the forests and lakes have always been subjected to the acid products of sulfur, why should we worry about a bit more? Estimates of the total amount of sulfur placed into the global air by human activity show that this amount is roughly equal to all other sources, and therefore the human impact cannot be described as adding "a bit more."[8] But a more graphic view of the human contribution emerges from studies of Greenland's ice cap.

Each year, snow falling on the permanent ice of Greenland (or Antarctica, for that matter) deposits particles from the air. Snows in subsequent years not only cover and preserve these particles, they also trap some air that is mixed among the snowflakes. Year after year the later snows compress the earlier falls into ever thinner layers and transform them into ice, without changing the composition of the trapped particles and air. Brought to the surface today and analyzed, layer by layer, this ice tells a fascinating story. Down to a considerable depth, the layers remain distinct enough to be easily counted, like tree rings, and scientists can readily determine when each sample was laid down. At greater depths, the compression of the ice makes dating annual events more difficult, but less precise techniques can be used to tell approximately how old the sample is. Substances originating in North America and Europe, and in even more distant places, are frequently deposited in northern polar regions. Thus, Greenland's ice stores a record of how activity in the Northern Hemisphere, which includes most of the world's industrial societies, influences the air. Figure 1, a graph of the amount of lead found at various depths in the Greenland ice cap, has little to do with acid rain, but it forcefully illustrates the record-keeping ability of this ice cap.[9] Note the steady climb in lead deposits, starting sometime before 1750, due to the expanding use of lead and the increasing number of lead smelters in Europe and North America. Lead occurs with silver in many deposits, so the increasing recovery of silver for coinage and

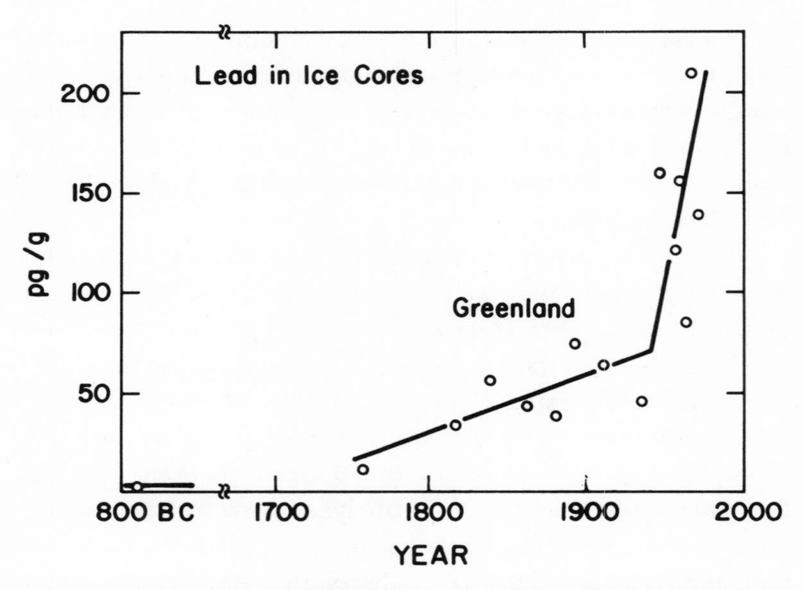

FIGURE 1. The amount of lead in samples of Greenland ice recovered from various distances below the surface. The units of lead concentrations in the ice are picograms (10^{-12} grams) per gram of ice, or equivalently, parts per trillion by weight.

jewelry also contributed to the rising amounts of lead in the ice. But once it was realized that the lead and other metals lost to the air could be economically recovered, and smelters became more efficient, the quantity of lead in the air did not increase as fast as the number of smelters would suggest. In the twentieth century, industrial activity continued to accelerate, and tetraethyl lead was introduced as an additive for gasoline. As a result, the amount of lead in the Greenland ice soared to more than two hundred times its pre-civilization values. In the future, scientists should be able to deduce from analyses of Greenland ice whether the current efforts in many industrial countries to reduce lead use in fuels were successful.

Another substance that can be measured in the Greenland ice cores is sulfur in the form of sulfate.[10] The ice record can be used to check the conclusions of scientists trying to estimate the amount of sulfur emissions derived from human activities. As figure 2 shows, sulfate concentrations also begin to increase rapidly sometime around 1800, and this trend continues in the

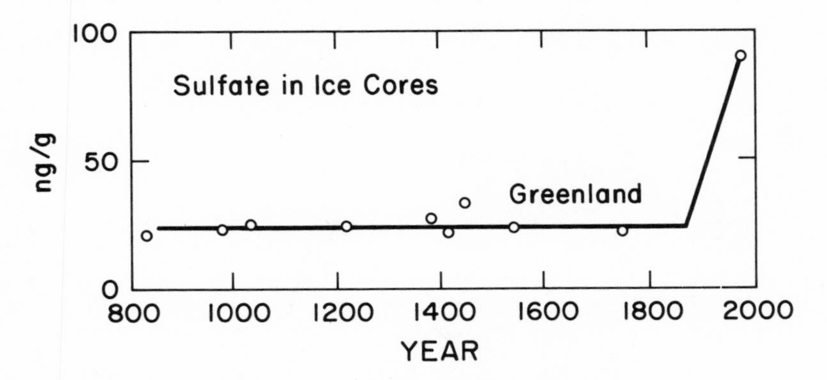

FIGURE 2. The concentration of sulfate ions in Greenland ice in nanograms sulfate (10^{-9} grams) per gram of ice, or parts per billion by weight. The line was drawn through the measured points to emphasize the rapid increase in recent years, but the actual time of the start of the rise, as determined by these few points, could have been any time after about 1750.

most recent samples, in which there is more than three times as much sulfate as in earlier times. Almost certainly, both the lead and sulfate amounts mirror the increasing industrialization of the Northern Hemisphere. Neither lead nor sulfate from smoke-stacks remains in the air for more than a week or two, and a large portion of these substances falls out before reaching Greenland. Thus, an increase of the Greenland deposits over the last two centuries represents an even larger increase elsewhere.

The measurements shown in figure 2 are those for which the ice samples showed no evidence of volcanic debris. Volcanoes can emit sulfur dioxide into the atmosphere, and they can dominate the sulfate in the Greenland ice for short periods of time. For example, figure 3 (for source of data, see n. 10) shows the sulfate in the ice at depths corresponding to a few years around A.D. 535 when an unknown volcano pushed the sulfate values to almost 200 nanograms per gram for a short time. Famous volcanic eruptions, such as that of Krakatoa in 1883 and Eldgja in 934 show clearly in the detailed analysis of the ice cores. These events, however, are sufficiently rare that they do not dominate the long-term average amounts of sulfate in the ice. What the Greenland

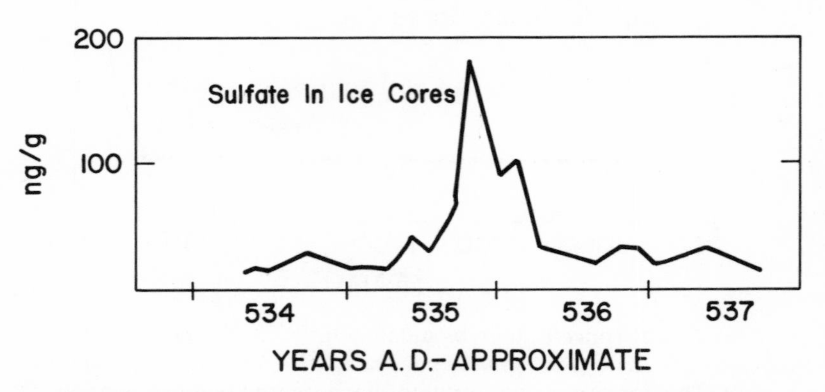

FIGURE 3. Sulfate measured in the Greenland ice at depths corresponding to four years near A.D. 535.

ice confirms is that human activities exceed nonhuman processes as sources of sulfur in the air. What it cannot tell us, of course, is what that means.

The present pattern of life on earth evolved with a certain rate of sulfur deposition, an amount typified by that falling on Greenland a few hundred years ago. It is reasonable to expect that if that rate triples, as it has in the Northern Hemisphere, many features of living systems must also change, even those systems located far from urban centers. Indeed, early alarms have already been sounded.

The modern acid-rain problem first came to public notice in the 1960s when fishermen began complaining that the number and kinds of fish available in some remote lakes had declined strikingly. Studies of these lakes, first in Scandinavia, then in North America, Scotland, and elsewhere, showed that the lakes had indeed changed: they had become so acidic that many aquatic species could no longer survive in them. Soon scientists were conjecturing that this change in acidity had been caused by sulfuric acid in the rainfall feeding the lakes and that the sulfuric acid originated in smokestacks emitting sulfur dioxide.

Scientists found it fairly easy to show that an acidified lake could no longer support some aquatic species. And it was easy enough to show that rainfall in Scandinavia and elsewhere contained sulfuric acid. But in order to connect the burning of fossil fuel with the death of fish, scientists had to establish that sulfuric acid in the rain acidified the lakes and that particular smokestacks put the sulfuric acid in the rain in the first place—a much harder task than demonstrating that acid kills fish.

Mapping the chemistry of even a single lake is not a simple proposition. Usually, most of the water entering a lake does not fall as rain or snow directly on the lake surface. Instead, the lake collects water as runoff or groundwater from a much wider catchment area. Rain falling far from a lake may strike leaves as it falls. When it hits soil, it may flow on the surface or soak in and move underground, contacting several kinds of soil before entering the lake. It may also flow across the surface of rocks. At each of these encounters it may undergo chemical reactions that make it more or less acidic and change the amounts of other impurities in it. Understanding a lake's chemistry thus requires studying and comparing many lakes, reading the layers of sediments at the bottom of the lake to determine when acidification began, and describing the nature of the chemical changes experienced by the rain that finally reaches the lake.

Such painstaking studies have been done. They have built an impressive chain of evidence connecting acid in rain to the deterioration of lakes, and many of the important, detailed processes involved are now understood.[11] For example, the studies were able to explain the puzzling observation that two lakes could respond to similar amounts of acid rainfall in quite different ways. The dissimilarity arises from differences in the soils and rocks near the lakes. If the lake's basin contains soil or rocks that can neutralize the acid in the rain, the lake may show little response, so far, to the acid rain. Meanwhile, a nearby lake surrounded with different material may already have lost its fish and other aquatic species. Lakes, or sometimes whole regions, can be characterized by their rocks and soils as "sensitive" to acid rain or as "well buffered." In short, the soil around some lakes has enough neutralizing power to slow or eliminate the ill effects of current rates of acid deposition.

All lakes would eventually become "sensitive" when the annual dose of acids had finally used up all of the available neutralizing capacity of their surrounding rocks and soil if no other means of enhancing their buffering capacity were at work. Fortunately, other processes tend to restore or replenish the acid-neutralizing capacity. One is the weathering of certain rocks,

which releases additional buffering material to the soil. Certain microorganisms can process sulfates and reduce their impact on lakes. Scientists also suspect that, in the United States, dust blowing in windstorms from alkali deserts in the West and landing on eastern watersheds may help neutralize acid rain.

These findings took years to develop. Meanwhile, researchers attempting to trace the origin of sulfuric acid rain faced an equally complex task. But in this case powerful circumstantial evidence pointed to smokestacks as the source. Estimates of the total amount of sulfur falling as sulfuric acid over a region such as eastern North America roughly equaled the total amount of sulfur emitted into the air in and upwind of that region. Adding to the case was the fact that the increase in acidic precipitation over the last few decades had occurred in regions where sulfur dioxide emissions had increased during the same time. And, finally, the areas with the most acidic rain were, on the average, in and downwind from areas emitting the most sulfur into the atmosphere. Still, circumstantial evidence never quite carries the day in scientific investigations, and filling in all the details between smokestack and rainfall with good measurements and convincing theory has proved expensive and time-consuming.

Some basic and not-so-basic chemistry has been involved in this sleuthing. Measurements in the air and in the laboratory were required to track down the processes by which sulfur dioxide was transformed into sulfuric acid. It turned out that there was no single path: this transformation could take place after sulfur dioxide was dissolved in a droplet of water in a cloud or earlier, in the clear air, in the presence of sunlight. In both reactions, other substances were involved, some of them also human-produced pollutants. How sulfuric acid or sulfates, once created, reached the surface of the earth was also studied, and here again multiple pathways were found. These compounds could fall to the ground in the raindrop in which they were formed, they could form in the dry air and then be washed out by a falling drop, or they could simply hit the ground as gas or a dry particle and attach themselves to a leaf or a bit of soil. This last process—so-called dry deposition—has proved much more difficult to monitor than the acids in rain or snow, and scientists still have only rough estimates of the total amount of dry acidic deposits.

Quite early in these more recent studies, scientists found that rain contained nitric acid as well as sulfuric acid. This discovery meant that two lines of investigation were required: all the measurements of sulfur dioxide and sulfuric acid had to be taken for nitrogen oxides and nitric acid as well. Not only the chemistry, but also the distribution of sources of the two substances differ. While most sulfur gets to the atmosphere during coal burning in large power plants or the smelting of metals such as copper, the sources of nitrogen oxides include the combustion of all fossil fuels, whether in power plants, industrial processes, or internal combustion engines—mostly cars, buses, and trucks. Thus the sources of nitrogen oxide are more widely dispersed than those of sulfur dioxide.

But nitric acid did not have the centuries-old bad reputation that airborne sulfur had, so some scientists were at first inclined to dismiss nitric acid as an important contributor to acid rain. After all, nitrates are plant food; farmers spend millions applying nitrogen-based fertilizers to their fields each year. It seemed reasonable to expect this kind of acid rain to do as much good as harm to forest and field and to be largely utilized by plants before it could flow to and contaminate a lake. But, unfortunately, this was not the case. In winter, acids accumulate in the snow; in the spring thaw they are suddenly released. Nitric acid proved to be the major contributor to this "acid pulse," which does much of the damage to lake ecosystems. In the western United States, where coal tends to have less sulfur than elsewhere, nitric acid contributes more than half the total acidity to rainwater. And nitrogen oxides were found to initiate chemical reactions in the air that generate ozone and hydrogen peroxide, both of which were implicated in the transformations of sulfur dioxide to sulfuric acid and both of which can damage plants directly.

Even when the sources of sulfur dioxide and nitrogen oxides are identified and the main chemical transformations are understood, a problem remains in describing how the substances move, while undergoing transformations, from the smokestack or automobile exhaust pipe to the watershed of a sensitive lake. Observing a single cloud of pollutants continuously for several days as it moves away from its source, changes, mixes with other pollutants, and finally is deposited on the earth's surface is usually not possible. Complex, computer-based, mathematical models of large regions are therefore used to simulate the process.

Describing a model as complex or mathematical causes many people to regard all models as mysterious, but some are in everyday use. A commuter driving a car in traffic has a mental model of where other cars will be in a few seconds and decides whether to speed up, slow down, or turn on the basis of this continuously changing projection. Only when the items to be considered are too numerous or variable for a person to remember, or when the rate of change becomes too fast for a person to project usefully, is a computer with its large memory and high speed brought in. An air-traffic controller has much the same problem to solve as the car driver, but the speeds are greater, the number of vehicles involved is larger, and the vehicles can move up and down as well as right and left, so the controller uses a computer to generate a picture of the activity and a continuous projection (or model) of future positions and potential collisions.

Such numerical models are common features of modern life. Large businesses design models to portray their main activities and use these ever-changing maps to help decide which activities to emphasize and which to decrease or terminate so as to earn the highest profits. Highway departments model traffic flow to decide where to place and how to time stoplights and where to recommend additional roads. The weather service prepares daily forecasts with the aid of extremely elaborate models of atmospheric motions. These models are quite different from each other, but all of them start with an observed situation: the activities of the business in a given year, the distribution of traffic on a typical morning, or the temperature, pressure, and winds at many places in the atmosphere over a large region. All these models also rely on mathematical formulas in order to calculate future behavior or events. For the atmosphere, the model designers use well-established physical laws relating forces and motions to provide the formulas. For highway flow, the model may be based on an observed relationship between the density of traffic and the speed with which people tend to drive.

The models used in acid-rain studies are as complex as any. They involve techniques for simulating a region's weather and methods for calculating the myriad potential interactions among dozens of the chemicals found in the real atmosphere. The range of possibilities that must be considered is large. A single power plant may emit sulfur dioxide, various nitrogen oxides, particles of soot and other substances; a single car can be the source of carbon monoxide, unburned hydrocarbons, nitrogen oxides, and particles of various composition. Sunlight striking this mix of substances can produce ozone, hydrogen peroxide, and the range of more complicated chemicals we associate with urban smog. And each of these chemicals can react with the others.

At one time it was hoped that model calculations would make the job of regulating smokestack emissions easier. If we could determine which smokestacks and tail pipes are contributing acids to sensitive lakes, we could focus our attention on the troublesome sources and not go to the expense and bother of regulating the others. But both models and measurements remind us that winds are not constant, so that over many years chemicals from a single source contribute to the acid in all the lakes over a wide region, and a single lake receives acids from many smokestacks in every direction. Furthermore, in the air the transformation of sulfur dioxide and nitrogen oxides into sulfates, nitrates, and acids is governed by such other pollutants as ozone and hydrogen peroxide, which in turn are generated by still other pollutants, such as nitrogen oxides and hydrocarbons. The models were essential to achieving a general understanding of this complex situation, but the message to would-be regulators was a tough one: You must find ways of decreasing emissions of many substances from all sources in a very large region.

The news was not much better from the scientists studying the varied ways in which sulfur and other compounds work their way through living systems. For example, it was demonstrated that fish in acidified lakes suffer direct damage from the acids to their gills and their bodies, but they also suffer indirect effects. When acidic rain passes through soil on the way to a lake, it dissolves aluminum that would ordinarily be tightly held by soil particles. This aluminum is carried to the lake, where it poisons some kinds of aquatic life. Acid in rain was shown to change the character of soils in other ways as well, to affect microorganisms that help maintain soil productivity, and to damage the small roots that plants use to acquire nutrients.

Plants differ widely in their responses to the sulfur dioxide that has not yet been transformed into sulfuric acid: some damage easily, others are more resistant. Given nature's survivalist tendencies, an ecosystem subjected to daily doses of sulfur dioxide can thus be expected to shift the mixture of its plant populations. Furthermore, sulfur compounds in the air react with ozone and hydrogen peroxide, each of which can harm living materials. We have come a long way from the intense clouds of sulfur dioxide at Copper Hill that directly killed trees and poisoned soils. Now each source creates a list of chemicals; each living thing experiences a list of insults.

Despite the rapid progress that had been made in understanding the complex interactions of sulfur and nitrogen oxides in the atmosphere and biological systems, much of the policy debate remained back at the starting line. So what if some person could not catch a fish? According to one U.S. government official, putting cleaning equipment on sulfur-producing smokestacks would cost six thousand dollars for each fish saved. Such narrow arguments have delayed for nearly a decade any concerted, national approach to reducing acid rain or any agreement with Canada to take mutually advantageous steps to clean the air. A more clear-headed appreciation of the breadth and complexity of the interaction of pollutants with living things and of the long-term increase of sulfur and nitrogen oxides in the global atmosphere might have led this official to regard the damaged fish as forerunners of more harm to come, as the tip of the iceberg. Indeed, only a few years were needed to show that such was indeed the case.

In the early 1980s, reports began to emerge from Germany that trees were dying unaccountably in the Black Forest. Other areas in Europe noted similar events. Disease, insects, drought, unusually hot or unusually cold weather were all suspected of causing the damage, but no convincing case could be made for any of them. The change was dramatic: in 1982 an estimated 5 percent of the trees in the Black Forest were damaged; by 1985, measurements indicated that 50 percent were.

Meanwhile, similar observations were being made in the eastern and southeastern United States. Studies of the tree rings within living forests showed that while in earlier decades fluctuations in tree growth could easily be related to years of favorable or unfavorable weather, more recent declines in growth could not. Now attention is being focused once again on the impact of harmful compounds in the air: direct damage to leaves and needles by sulfur dioxide, hydrogen peroxide, and particularly ozone; the harm that acids in the soil do to roots; the effects of aluminum released in the soil by acids; the impact of overfertilization caused by nitrates arriving through the air; and the reaction of trees weakened by these combined events to recurring extremes of weather and normal attacks of insects, bacteria, and fungi. These recent studies have also revealed that water droplets in fog or clouds can be much more acidic than the water in rain and that these tiny drops, some as powerful as battery acid, can cause damage to leaves and needles beyond that created by exposure to pollutant gases. This discovery helped explain why trees at higher altitudes, which are frequently bathed in clouds, appeared particularly susceptible to damage. Thus, the death of fish in Swedish and Canadian lakes proved to be only the earliest signals of the sweeping changes in natural ecosystems being wrought by industrial emissions.

All of these recent discoveries make the design of a cure more difficult as well as more necessary. Because several pollutants are responsible, no single control strategy will be sufficient. Because several atmospheric and biological pathways are involved, no single treatment of lakes or forests will heal the damage. And because important sectors of our industrial society create the pollutants, changes will be resisted.

One of the first steps in proposing a cure is to decide how great a reduction in pollutant emissions is required. It is unlikely that emissions of harmful substances can be reduced to zero; it is also clear that today's emissions are excessive. At what level between these extremes should we aim?

We can guess that plants and ecosystems in general are adapted to live with amounts of sulfates deposited by precivilization sources. So if we could limit emissions to a 5 or 10 percent increase over those deposits, we would probably do no damage.

The Greenland measurements suggest that over vast areas far from industrial sources of sulfur, the biosphere should be able to withstand a bit more than a third of current rates of deposition. But to get the Greenland deposition down to such an amount would require a 95 percent reduction in Northern Hemisphere emissions—an unlikely achievement.

Another approach is to ask botanists and ecologists to answer a different question: What rate of deposition can be tolerated by most ecosystems in most locations? For sulfur, they conclude, limiting deposition to 17 kilograms of sulfate per hectare per year will preclude obvious damage to all but the most sensitive aquatic ecosystems. Other researchers are more cautious and recommend reducing deposits to 10 kilograms per hectare per year, compared to annual depositions today of from 20 to 50 kilograms per hectare per year in the eastern United States and Western Europe. (For the record, regulations have already prevented much larger sulfur emissions and hence larger rates of deposition.) These findings and recommendations derive exclusively from extensive studies of lakes and fish—locations where acid-rain damage has been studied the longest—and not from damaged forests, where causes and effects are more complex. These proposed limits also do not apply to nitric acid, whose interactions are less well understood.

Whatever exact goal is set, the first step in reducing sulfur levels in the air is simple: Use less coal. This step requires neither great austerity nor nuclear power, merely the recognition that much of our coal-generated electric power is used inefficiently and that by improving energy efficiency we can reduce atmospheric problems and save money at the same time. Reducing the use of fossil fuels is part of a larger strategy to solve larger atmospheric problems, and will be dealt with in more detail in chapter 7. In the meantime, the problem of acid rain illustrates just how closely intertwined the atmosphere and the biosphere are and how seemingly minor changes in the atmosphere can produce complex damage to living things.

Inevitably, we will eventually be forced to consider the total biospheric impact of the substances we—as consumers, as drivers, as manufacturers—place in the air. But at present the acid-rain debate is still largely concerned with the damage to lakes in the northeastern United States and eastern Canada being caused by sulfur, various corrective measures, and who should pay for them. But whether the policy debate focuses on the short term and regional damages or the broader interactions now known to take place, the answer is the same: all the evidence indicates that we must greatly reduce sulfur and nitrogen oxide emissions to the atmosphere.

It is also sobering to realize that whatever people and nations decide to do about sulfur in the air, the record of their decision will be stored, for thousands of years to come, in the snows of Greenland.

3. Stratospheric Ozone

Damage to the layer of ozone in the high atmosphere by human activity is complex, esoteric, and completely invisible to anyone but the scientists who are studying the issue. Yet, around the world, people who twenty years ago had never heard the word *ozone* are now worried about its disappearance.

Ozone in the high atmosphere plays an important role with respect to life on earth and the structure of the atmosphere. When the intense sunlight reaching the high atmosphere breaks oxygen molecules into two oxygen atoms, most of these atoms reassemble, not as common oxygen, but as a molecule of ozone containing three oxygen atoms. Ozone is destroyed by being turned back into oxygen in a different set of reactions.

Ozone's structure allows it to absorb a certain kind of ultraviolet sunlight that would otherwise reach the surface of the earth and affect living material. The radiation of most concern is usually called UV-B and includes light between the wavelengths of 280 and 320 nanometers. Longer wavelengths reach deeper into the atmosphere, but are much less potent in producing biological changes. Shorter wavelengths are almost completely absorbed in the atmosphere and can therefore have little biological impact. Thus studies of the interaction of ultraviolet sunlight and living things focus on UV-B. This radiation can cause sunburn and certain skin cancers, reduce soybean yields, and damage surface-dwelling fish. The effectiveness of this radiation in changing biological material suggests that almost any living tissue exposed to it suffers some effect.[12]

Ozone plays an important role in the high atmosphere in addition to screening out UV-B. By absorbing ultraviolet sunlight, ozone deposits the heat associated with this light into that level of the atmosphere, thus creating a layer much warmer than those immediately below. The stable region so created is the stratosphere. It is in this stable layer that disturbing changes are occurring. As scientists' understanding of the chemical reactions that create and destroy ozone increased, it became clear that relatively small quantities of some substances could change these reactions and hence the amount of ozone in the stratosphere, provided those substances were placed in the high atmosphere. And chlorine, an effective chemical catalyst that can change ozone into normal oxygen, is appearing in rapidly increasing concentrations in the stratosphere.

Some chemical reactions require a catalyst, a substance that is essential for the reaction to proceed rapidly but that is not consumed during the reaction. Catalysts abound in modern industrial chemistry. In the United States today, many new automobiles are built with "catalytic converters" in the exhaust lines to cause the carbon monoxide in the exhaust, or the last bit of hydrocarbons remaining unburned, to be converted into carbon dioxide and water. The catalyst in this case is a screen containing precious metals on whose surface the reaction proceeds many times faster than it would otherwise. Catalysts can be extremely efficient. It is not uncommon in modern chemical engineering practice for an atom or molecule of a catalyst to pass through the reaction ten or a hundred thousand times before an unintended side-reaction removes it permanently from the process.

Each day, ozone is created during daylight hours by reactions driven by intense sunlight. Each day, a fraction of all the ozone in the stratosphere is destroyed by reactions with chemicals occurring naturally in the stratosphere. The amount created is more or less fixed, while the amount destroyed increases as the total amount of ozone increases. The amount of ozone builds up until the amount created equals the amount destroyed and an approximate equilibrium is reached. If, however, we introduce into the stratosphere a new substance, such as chlorine, which catalyzes the destruction of ozone but not its creation, a new equilibrium will have to be reached in which there is less ozone than before. And if the new substance is a very efficient catalyst, as chlorine is, then a very small amount of it can produce important changes in the ozone layer.

The equilibrium between the creation and destruction of ozone is an example of a common phenomenon in geophysics, but one that seems to puzzle many nonscientists. An illustrative analogue may clarify the confusion.

Suppose you have an empty steel barrel and you pour in buckets of water at the rate of one gallon per minute. Suppose in addition the barrel has, up one side, a row of small holes, such as nail holes. After the first gallon of water is dumped into the barrel, the lowest hole begins to leak, but the leak is far too slow to get rid of a whole gallon before the next bucket is poured in. In this way the level of water in the barrel will rise with each new addition. But as the level of water rises, the leaks get stronger, both because there are more holes leaking and because the water pressure on the lower holes is greater. Finally a point is reached at which the leaks total one gallon per minute. At that moment the level of water in the barrel stops rising: the amount going in each minute equals the amount leaking out—equilibrium has been reached. This equilibrium lasts as long as you keep pouring in water at a gallon per minute and nothing changes about the holes.

The two features of this down-to-earth illustration that are similar to the atmospheric cases discussed in this book are the steady input—the one gallon per minute—and the output that depends on how much water has accumulated—the leaks that increase as the water level rises. Stratospheric ozone has these two features. The amount of ozone put into the stratosphere each day is steady—it depends almost entirely on the intense sunlight at that altitude and does not change much from day to day. The loss of ozone, however, depends on how much ozone is available for the various reactions to destroy. In this manner, an equilibrium is reached.

To push the analogy between the barrel and the stratosphere a bit further, adding chemical catalysts to the stratosphere is the same as punching a few more holes in the barrel while continuing to pour in the same one gallon per minute. The water in the barrel will settle down to a lower level, a new equilibrium, in which once again the leaks are just one gallon per minute. So too, the addition of catalysts to the stratosphere will lower the total amount of ozone.

We now have the three links of a long, logical sequence. Chlorine in the stratosphere reduces the amount of ozone in the stratosphere. Less ozone in the stratosphere means that more ultraviolet sunlight penetrates the atmosphere. Increases in ultraviolet light cause more damage to living things at the earth's surface. Another line of reasoning, relating to ozone's role in defining the stratosphere, has received less attention from scientists. Chlorine in the stratosphere depletes ozone, more at some altitudes than at others. A change in the distribution of stratospheric ozone affects the height above the earth at which the stratosphere is most strongly heated by absorbed sunlight. Changes in the heating pattern of the stratosphere modify the winds that blow at these altitudes, thus altering the stratospheric climate and further changing the distribution of ozone.

Ordinarily there is very little chlorine in the stratosphere. Chlorine gas is sometimes spilled in industrial or shipping accidents, but this gas reacts strongly with almost any waterdrop or particle it touches and as a result is used up long before it can diffuse upward. Ocean waves throw up small droplets of salty water, some of which evaporate, leaving salt particles in the air. Although these particles contain chlorine, the chance that one of them will get as high in the atmosphere as the ozone layer is small, since salt is very soluble and these particles are readily washed out of the air by the rain. Some biological systems emit methyl chloride, a gas that contains chlorine. But this gas reacts fairly rapidly with other substances, and most of it disappears before it can diffuse to the stratosphere. Thus, strong barriers prevent chlorine from reaching high in the atmosphere, unless people contrive to put it there.[13]

If we did wish, for some reason, for chlorine at the earth's surface to move into the stratosphere, we would have to arrange for the emission at the surface of the earth of a chlorine-containing gas. We would, in addition, have to find a chlorine-containing gas that did not react readily with anything, one that was not very soluble, and one that, upon reaching the stratosphere, could be broken down to release free chlorine only by the action of strong ultraviolet light. (If it were broken down too soon, by sunlight that penetrates low into the atmosphere, the free chlorine would react with something and be removed.) The properties I have just described would also make the gas extremely useful here at the surface of the earth, and people have worked hard to create just such a substance.

If a gas does not react with other substances, it is less likely to be toxic to people who accidentally breathe it, and it is not likely to corrode pipes or vessels in which it is carried or stored. It could be used, for example, as a compressed gas in spray cans to help dispense hair spray or other substances without affecting those substances. It could be used to fill up the small cells in foam insulation without reacting with the plastic or harming the people who use the foam, or it could simply be compressed and used to blow dust off photographic lenses or delicate electronics equipment without scratching the object or endangering the user. If, in addition, we could liquefy the gas at reasonable pressures and temperatures, it could be used to move heat from inside refrigerators and air conditioners outside to be dispersed. This could occur without danger of corrosion in the equipment and without risk to the people using the devices, even if the gas were to leak into the room or automobile being cooled.

Laboratory chemists created such substances decades ago. They are called chlorofluorocarbons, indicating that they contain carbon, fluorine, chlorine, and sometimes hydrogen. The name is frequently abbreviated to CFC, and a numbering scheme is used to tell how much of each element is in the molecule of the particular CFC under discussion. CFC-12, for example, has one atom of carbon, no atoms of hydrogen, two atoms of fluorine (and, by implication, two atoms of chlorine) in each molecule.

Two of these substances, CFC-11 and CFC-12, have proved so valuable in a number of applications that more than 20 million tons have been manufactured worldwide. Most of this 20 million tons still exists and either has escaped to the atmosphere or eventually will. Once in the air, these substances mix and diffuse, finally reaching all parts of the atmosphere. Those CFC molecules that find themselves in the stratosphere are subjected to intense ultraviolet radiation from the sun; they split apart into smaller fragments, releasing chlorine. The chlorine then starts a new career as a catalyst in the reactions that destroy ozone.

This brief description clearly indicates that there is cause for concern, but it does not demonstrate that we indeed have a problem. That demonstration depends on hard data: how much CFC is being released, how much gets to the stratosphere, how much ozone is depleted by that much CFC, and how much damage will result from the extra UV-B admitted and where.

The first question can easily be answered. Figure 4 shows a series of measurements of the concentrations of CFC-11 and CFC-12 in the atmosphere.[14] From these numbers we can estimate how much chlorine there is in the stratosphere today and

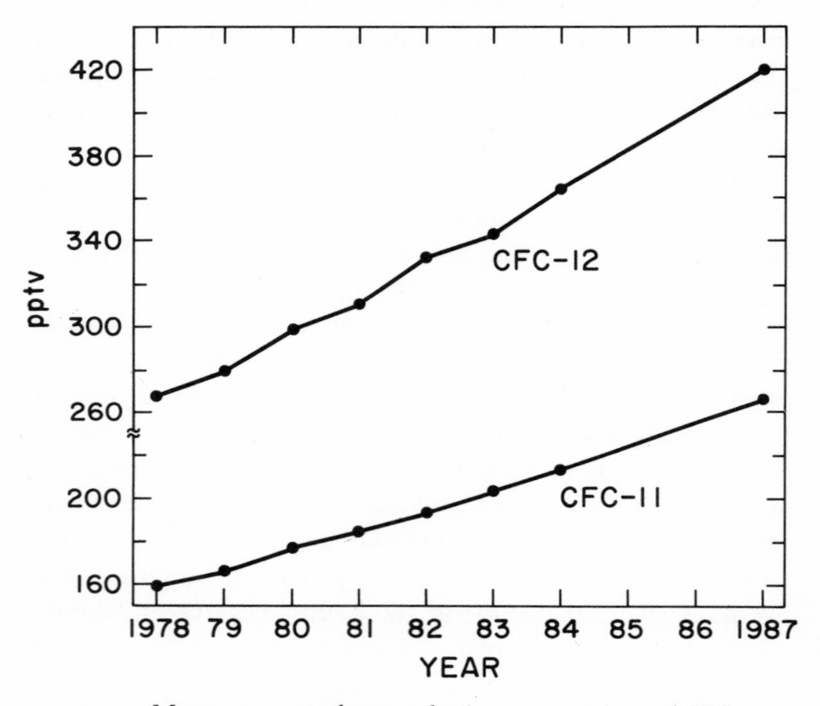

FIGURE 4. Measurements of atmospheric concentrations of CFC-11 and CFC-12 made at Ragged Point, Barbados. The concentrations are plotted in units of parts per trillion by volume. These chemicals are monitored at a number of locations around the world, and the results are similar, but not identical, at all sites. Measurements made in the Southern Hemisphere, for example, are slightly lower than those made farther north, indicating that some time is required for these gases to diffuse from the regions of high use in the industrialized countries to other parts of the world.

how much is likely to be there in the near future. The graphs show that the concentrations of these two substances are increasing very rapidly, more than 5 percent per year. (For purposes of comparison, 5 percent per year is three times faster than the world's "exploding population" is growing). The amount of carbon tetrachloride in the air is increasing at about 1 percent a year and it too can survive to reach the stratosphere and there produce free chlorine. Chemists also realized that chlorine was not alone in its ability to catalyze ozone destruction: bromine is even more effective and is beginning to be detected in the air as part of a family of substances called halons, which were finding expanded use in fire extinguishers.

The other questions are more difficult to answer because stratospheric changes are not purely chemical. For example, ozone is found over each pole of the earth during the long polar night, despite the absence of sunlight, because stratospheric winds transport ozone into the polar regions. Furthermore, as mentioned in the second "logical train of possible harm," when a part of the stratosphere is illuminated by sunlight, ozone absorbs most of the energy that goes toward warming the stratosphere and so driving the winds. Thus, it is necessary to understand the relationships between and among ozone, sunlight, winds, the chemicals that have always been there, and the chemicals now being added to the stratosphere by the activities of people.

The path to a better understanding of these complex interactions leads to a mixture of activities typical of scientific research today: measurements in the atmosphere (how much of what is where, and how is it moving?) and in the laboratory (how rapidly does A react with B and what kind of light does B absorb?) and construction of some technique to synthesize the theory and measurements. As with studies of sulfur transport and acid rain, the technique required to synthesize what we know about stratospheric chemistry is numerical modeling.

To calculate how much ozone will be destroyed by a certain amount of chlorine in the stratosphere, a scientist could simply take the results of laboratory measurements on the catalytic efficiency of chlorine, combine that with measurements of the amount of chlorine in the stratosphere, and estimate the ozone loss. The answer would, however, be little better than a guess, since such a calculation would not allow for other important processes. For example, if some ozone is destroyed, more UV-B will be able to reach lower in the atmosphere. There it will trigger chemical reactions, some of which create new ozone. Which will be larger, the amount of ozone destroyed or the amount of ozone created? Or another example: the number of times a chlorine atom can serve to catalyze an ozone destruction depends on other substances in the stratosphere that remove chlorine—water and nitrogen compounds—that are, like ozone, moved around by the stratospheric winds. How much of each of these substances will there be at different altitudes above any particular place? Estimating the effect of each of these processes, and many more, soon leads to bewildering complexity, since each process, in addition to being complicated on its own terms, must be calculated in company with the others, not separately. The solution to this difficulty is to use the large memories and rapid calculating ability of modern computers. Without numerical models, there would be little hope of being able to estimate with confidence the effects of CFCs on stratospheric ozone.

Scientists have designed such models and have gradually increased their complexity as new features of the ozone problem have been recognized, and the numbers used to describe each process have been progressively refined by laboratory measurements. The current level of understanding and modeling indicates that if the manufacture and release of CFCs continue at the current rates, the total ozone in the stratosphere will decrease by several percent. It is also estimated that for each percent decrease in total ozone, there will be 2 percent more UV-B at the surface and 4 percent more cases of human skin cancer.

The complexity forced upon the stratospheric models and on the discussions of possible harm shows up in even the simplest measurements of ozone. Daily measurement (fig. 5) of the amount of ozone in the stratosphere above a single location, in

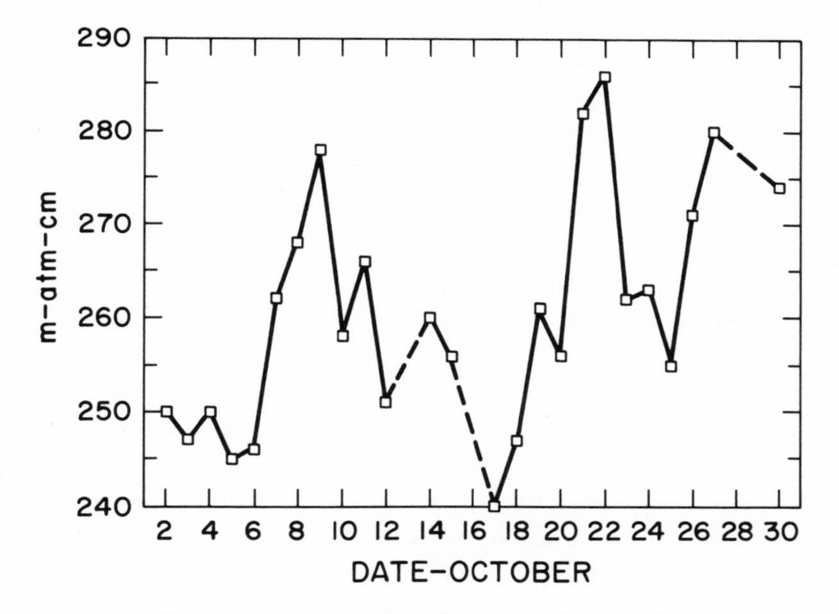

FIGURE 5. Measurements of the total amount of ozone above an observing station in Arosa, Switzerland. The units used for these measurements are milli-atmospheres-centimeter, which describe how thick a layer the ozone over the location would make if it were compressed to sea-level pressure and put at a temperature of 0° C. For example, 250 units means a layer 0.25 centimeter thick. These measurements show typical day-to-day variations. Ozone also varies throughout the year (maximum in the spring, lower in the fall, for mid- and high-latitude locations), and there are additional variations from year to year. The graph shows daily measurements connected by a solid line. The dashed line connects measurements where intervening days are missing.

this case a mountaintop in Switzerland, provides a case in point.[15] The first thing one notices is that the amount of ozone is quite variable (during the month shown it fluctuated by 17 percent), and so a decrease of a few percent in the annual average may be hard to detect. Another thought comes to mind when we look at these measurements: Plants and animals have continued to survive in Switzerland, so they can obviously survive hours or days of low ozone and hence higher-than-average UV-B. If a change in ozone of a few percent is going to be obscured by larger natural changes, and if living things can survive wide fluctuations in the amount of ozone, why worry about a little extra UV-B?

Even with sizable fluctuations, there are at least two ways in which an increase in the average ultraviolet radiation at the surface of the earth could damage living material: if the biological reactions respond to the cumulative dose, or if organisms respond to the extreme events. People who spend a lot of time outdoors, for example, tend to get skin cancers more often than other people, and these cancers occur on parts of the body, such as the face, that are normally exposed to the sun. This observation can be explained by supposing that the skin must receive a certain total dose of sunlight, perhaps over many years, before the cancer is induced. If this explanation is correct, then an increase in the average amount of ultraviolet light arriving at the surface, even if that increase is smaller than the usual fluctuations, will shorten the time required to accumulate the critical dose of sunlight and hence increase the occurrence of skin cancer.

Perhaps, instead, skin cancers appear when the skin is subjected for a short time to the intense ultraviolet radiation that arrives on those days during which the ozone is at the lowest level it reaches during the year. In this case, what matters is the probability that the ozone will be very low on the day a person is outdoors with skin exposed. A decrease in the average ozone (once again, even if the change is smaller than normal fluctuations) can change, perhaps dramatically, the probability of an extreme event, one in which the ultraviolet radiation is much stronger than usual. Figure 6 illustrates this point: as the average amount of ozone, and hence the intensity of UV-B, changes by a small amount, the number of low points, allowing the "harm" level of UV-B to be exceeded, increases greatly.

The direct impact on people of this projected increase in the average amount of UV-B radiation cannot be calculated, since scientists do not know to what extent people will change their behavior. In winter, people wear warm coats to protect themselves from the cold; perhaps in times to come people will wear hats, sunglasses, and long sleeves to protect themselves from the UV-B. Even secondary impacts could be modified by human action. Suppose that the amount of UV-B reaching the earth's surface by the year 2030 is too great for the economic production of soybeans. Agricultural groups have a long history of converting from one crop to another as conditions, both climate and economic, change. Therefore, the agricultural impact may be smaller than we might guess. The main blow might fall on the unmanaged biosphere—forest and grassland, phytoplankton and surface-feeding fish. Each of these ecosystems has evolved in a manner that allows it to tolerate the historical amounts of UV-B. An increase in that amount must produce changes: the replacement of sensitive species with more resistant species, lowered growth rates of all species, and change in the behavior of mobile species. The past history of the earth indicates that given thousands or millions of years, new species would evolve and new ecosystems would develop to restore the productivity of the biosphere despite the added UV-B. But with a change in conditions that is expected during the next few decades, evolution cannot

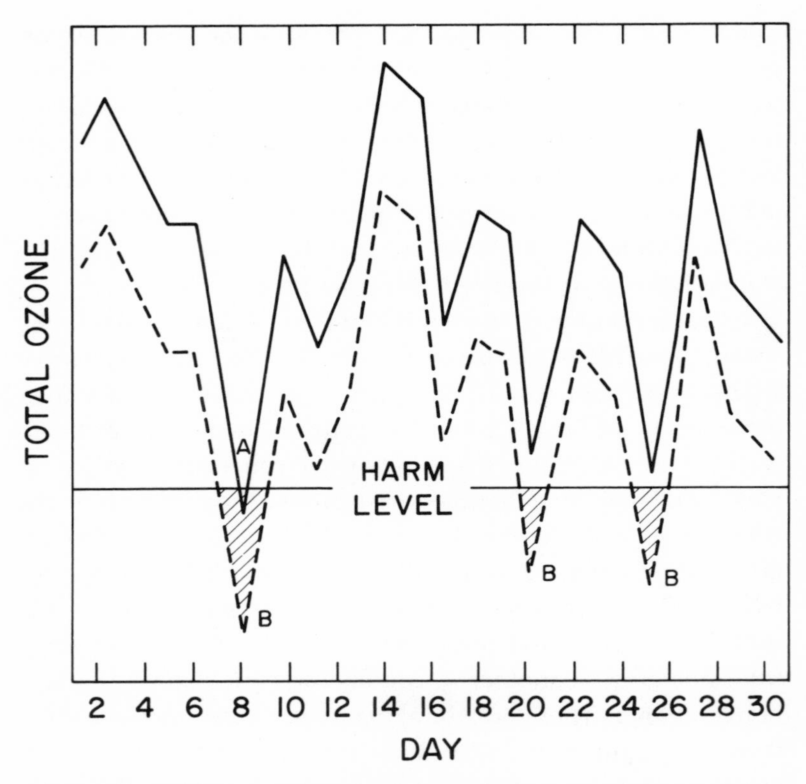

FIGURE 6. The effect of a small decrease in the average amount of ozone on the frequency with which the ozone level drops below some "harm" level. The solid line is a hypothetical set of daily values of total ozone above some location, similar to the actual measurements in figure 5. Once, during this time, at A, the value dips below a harm level. The dashed line is the curve that would result if all ozone amounts had been reduced by 3 percent by the action of stratospheric pollutants. The number of times the harm level is reached jumps from one to three (shown at B) and remains there for longer periods of time. This happens even though the decrease in the average, 3 percent, is much smaller than the usual fluctuations.

help, and the burden of adapting to the new environment falls to people and their governments. If this task is to be accomplished, the rate at which the changes are happening must be slowed.[16]

A start has been made. Representatives from most of the CFC-manufacturing countries agreed in 1985 that CFCs might be harmful and that they would exchange data and encourage research. This agreement was the culmination of several years of effort by the staff of the United Nations Environmental Program to create some limits to the production of chemicals that harm the ozone layer. The agreement created no limits but at least opened channels of communication among the nations. Then came a dramatic discovery. For a month each year, starting in the late 1970s, the total amount of ozone in the stratosphere over the Antarctic continent had been decreasing sharply and by a larger amount each year. It took some time for this phenomenon to be recognized, but by 1987 it was unmistakable. The total ozone over some locations had dropped 60 percent during the Southern Hemisphere's spring; at some altitudes the ozone was almost completely destroyed.

This Antarctic springtime decrease had first been noticed by scientists at a ground station in Antarctica, where they were measuring the total amount of ozone overhead. An examination of observations from orbiting satellites not only confirmed the amount of the decrease but also showed that it was a widespread phenomenon, not just a localized occurrence near the ground station. Since ozone depletion by CFCs was being widely discussed at the time, these chemicals were immediately suspected of causing the "ozone hole," and scientists were urged to confirm or deny that such was the case.

The scientific community was pleased by the renewed opportunity to study the stratosphere, but there were uncomfortable aspects to the situation. Although many calculations and model simulations indicated that ozone should decrease as CFCs increased, none showed any reason for the decrease to be strongest over the South Pole. None projected the decrease to appear as an intense, one-month episode. None projected a decrease as large as those observed. Clearly, further research remained to be done. It took two expeditions to high southern latitudes; measurements in Antarctica from aircraft, balloons, and satellites and from the ground; and much rethinking of the details of stratospheric interactions to explain the special features of the region around the South Pole that were intensifying the ozone-destroying power of chlorine. The stratospheric air over Antarctica is colder than elsewhere, allowing ice crystals to form even though the amount of water is very low. In addition, the circulation of air around the Antarctic continent—which takes place entirely over oceans, unlike the circulation in the Northern Hemisphere—is quite circular and regular, and it serves to discourage mixing of Antarctic air with that from lower latitudes. Thus, the Antarctic stratosphere can store chemicals through the long polar night. And on the surface of ice crystals, and perhaps inside them also, these chemicals can react more readily than if they were free gases. The coming of spring and sunlight then releases a torrent of ozone destruction.

While this renewed interest in the stratosphere was producing a variety of active measurements and vigorous theoretical discussions among scientists, other events were under way which produced a dramatic conclusion.

The earlier agreement to encourage research and exchange data on stratospheric ozone also included provisions for further negotiations to consider the subject of how to limit emissions of CFCs into the air. Renewed negotiations were in progress as the scientific excitement about the ozone hole was building up. Although the negotiators felt that too little was known about the Antarctic problem to include it in their deliberations, publicity about the ozone hole was intensifying. Some people called for bans on the use of products containing CFCs, but, most important, American companies manufacturing these chemicals began to believe that an eventual phaseout of CFC use was likely and supported an agreement to achieve such a move. With portions of the industry willing to consider some limits, and with the ozone hole in every newspaper, the negotiators promptly achieved an agreement—the Montreal Protocol—to freeze or scale down the production of those CFCs and halons that had the potential to destroy stratospheric ozone.

The details of this agreement reveal what hurdles had to be overcome to achieve the required consensus. Each industrial country entered the negotiations with two desires: to reduce CFCs in the air enough to prevent a catastrophe and to emerge from the negotiations with much of its own CFC production intact, because making and using CFCs are profitable enterprises. Each developing country entered the negotiating sessions with a similar desire to avert a tragedy, but also with strong sense that it would be unfair to limit or ban a profitable activity before the Third World had had a chance to participate. In addition, each negotiating government had some relationship with and policy toward the CFC industries within its borders, and some countries were part of larger entities such as the European Community that put forward positions concerning controls. The resulting agreement therefore is a complicated compromise. It distinguishes between developed and developing countries, specific differences are allowed between specific developed countries, and the rate of phasedown is slower than most stratospheric experts had hoped for. The treaty went into effect on January 1, 1989, with signatories from most of the industrial world, but without the participation of two giant developing countries, India and China.

It is not possible to say exactly what the treaty's effect will be.[17] The most optimistic estimates come from those who remember that the use of CFCs in aerosol spray cans in the United

States dropped sharply in the 1970s, well before there was any regulation requiring a phasedown. Industry does not have much enthusiasm for products whose sales must decrease, so once a phasedown appeared likely, replacements for CFC spray cans were not only quickly introduced but heavily stressed in advertisements as "new" and "improved." At the other end of the spectrum, analysts who have studied the Protocol and understand its terms estimate that the world's production of CFCs by the year 2009, when all the cutbacks are in place, could range from somewhat more than half the 1986 level to 20 percent more than the 1986 level, depending on how many countries eventually join in the treaty and how certain terms are interpreted—terms that were intentionally left vague in order to secure broader agreement. With production anywhere in this range, the concentration in the atmosphere will continue to climb for many years, and we can expect other dramatic episodes like the "ozone hole." The harsh fact is that the present concentration of CFCs in the air is enough to cause trouble—most visibly over Antarctica, but occurring everywhere—and any increase in that concentration will only add to the problems we face.

Even if global production of CFCs ceased today, the atmospheric concentration would continue to rise for a few years as the gases stored in foams and refrigerators gradually leaked into the air. Then, over the next hundred years, the concentration would slowly decline to about half the peak value as the CFCs were being broken down in the stratosphere. Ozone would continue to be destroyed by chlorine released in this process, and we could expect several decades of ozone at levels equal to or lower than today's already depressed amounts. This instantaneous production shutdown would be the most rapid cure for the stratospheric ozone problem that could be imagined, though not one that could be realistically implemented because of the almost complete dependence of refrigeration, worldwide, on CFCs. If industry does not move quickly to adopt other ways of accomplishing the tasks now involving CFCs, pressure is likely to mount from some governments to invoke the provision of the Montreal Protocol that permits a reconsideration of the phase-down rate if evidence warrants. Such a reconsideration would involve an attempt to gain support from other countries for both a more rapid decrease in production and a lower final level of CFCs.

Despite the bleak prognosis, the fact that the nations agreed to take some action before there had been any demonstrated harm to people or other living things is remarkable. Scientists have for decades found numerical models to be useful tools in their investigations of the complex world we live in, but convincing national leaders and administrators to take unpleasant actions on the basis of projections made with these models has been difficult. In the interplay of factors that shape political decisions, a scientific calculation resulting from a mathematical model is not a powerful force. The scientist who presents the calculation is eager to emphasize the limitations inherent in the particular approach taken, the nature of the approximations made, and the possibility that some important process has not been correctly included. It has therefore been easy for those who might be immediately disadvantaged by the actions suggested to label the projections as "just a model" or "only a theory" and to call for further study instead of action. Perhaps the "ozone hole" has made it possible for us to turn a corner in the application of scientific projections to public policy. The real test will come, however, in our reaction to climate heating.

4. Climate Heating

THE CHANGING COMPOSITION OF THE ATMOSPHERE

The current generation of scientists is being allowed to watch as a new force arises and grows—a force that is overtaking in importance the geophysical and astronomical causes of change in the atmosphere, and that is appearing rapidly enough to permit them to see some of its early results.

The realization that atmospheric science had to be broadened to encompass a new influence came slowly; scientists are no better than others at realizing when they are at an important turning point. But in retrospect, as always, one can now recognize those farsighted individuals who saw what was happening. My favorite seer is Hans Suess, a scientist now living in California, who more than thirty years ago was examining a rather minor scientific issue—radiocarbon dating—and in the process came across a surprising fact. Radiocarbon dating is a technique for determining the age of anything that was once living material. Willard Libby, a physicist, received a Nobel Prize for inventing this technique, and it has been used routinely ever since by archaeologists and others.

Some explanation of the details of the technique is necessary to understand the observation that caught Suess's attention. Cosmic rays from outer space enter the atmosphere and strike atoms that make up the air. These collisions sometimes result in the creation of a radioactive form of carbon—carbon 14, or ^{14}C—which is a bit heavier than ordinary carbon but chemically almost identical. This process has been going on for a very long time, so ^{14}C is in approximate equilibrium in the air: the same amount of ^{14}C is created each year as is removed. The interesting thing for archaeologists is that people and trees and grass are all part of that equilibrium. Plants pick up ^{14}C when they take in carbon dioxide as food, a bit of which has ^{14}C in it instead of regular carbon. When people and other animals eat those plants, they also pick up ^{14}C. In each case, the fraction of the ingested carbon that is radioactive is the same as the fraction in the air. The same is true for algae, snails, and the outer layers of trees. That fraction stays the same, even as we get older, because it is renewed each day by what we eat or breathe.

When we die, however, or the tree dies or the leaf drops, the renewal stops. The fraction begins to change, because the radioactive carbon gradually decays into something else while the regular carbon stays the same. It is as if the death of a living thing starts a clock that then runs on for thousands of years.

Figure 7 shows a simple picture of how to use that clock.[18] If a piece of wood is cut, at that moment it has its full amount of ^{14}C, shown as "1" in the figure. After a thousand years some of the radioactive carbon has decayed, and the amount is down to less than 90 percent of what it started with. After 5,730 years, the so-called half-life of ^{14}C, the amount is down to just half the original value. If an archaeologist finds a bit of charcoal buried in a particular layer at the bottom of a pond and wants to know how long it has been there, the ratio of radioactive carbon to ordinary carbon in the charcoal can be plotted on the vertical scale of figure 7. If, for example, the ratio is found to be 0.52 of the original amount,

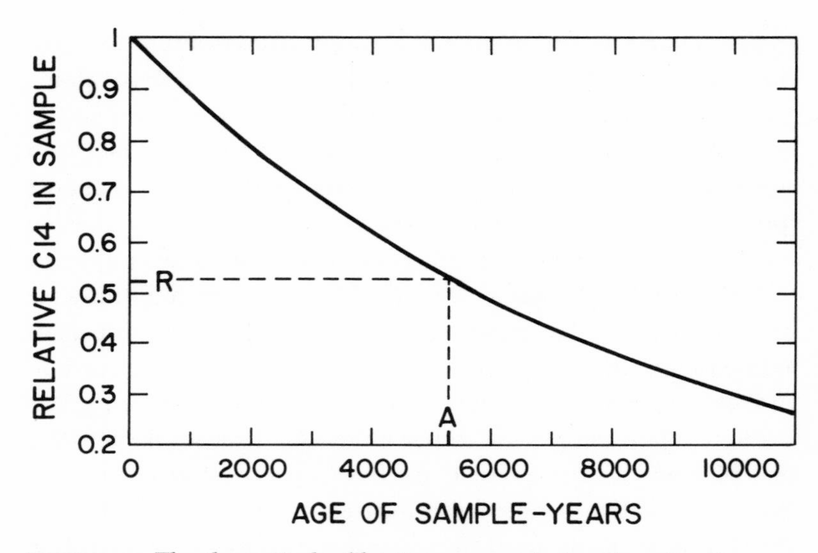

FIGURE 7. The theoretical calibration curve relating the ratio of radiocarbon to ordinary carbon in a sample to the age of the sample. A value of 1 would mean that ratio in the sample was the same as the ratio in the atmosphere.

plotted at R, an age of 5,400 years, shown at A, is deduced. The archaeologist can conclude that the wood of the charcoal lived 5,400 years ago and that the other material at the same layer in the mud at the bottom of the pond is likely to be just as old. A useful system, indeed.

But not perfect. The line in figure 7 comes from the simple theory of radioactive decay, which is undoubtedly correct, but when it is used to determine the age of a sample, one not only trusts that radioactive decay rates are always the same but also assumes that there has always been the same fraction of ^{14}C in the atmosphere so that all organic material, of whatever age, started with the same ratio. This assumption seems safe enough; in using this method to determine age, we are making use of some grand forces of nature. Cosmic rays come from the whole Milky

Way; the number reaching the vicinity of the earth is adjusted by the sun and its vast array of magnetic fields and streams of particles, and the number allowed to strike the atmosphere is controlled by the magnetic field of the earth, which stretches far out into space. We would not expect the amount of ^{14}C in the atmosphere to have changed rapidly by large amounts during times of interest to archaeologists, and in fact it has not.

But there are measurable changes, so that the actual curve used to interpret samples is not quite as regular as plotted in figure 7—it has a number of small wiggles in it. Both the sun's and the earth's magnetic fields change from time to time, giving us years with a slightly higher production of ^{14}C and years with slightly less. Thus, in practice one must use a more precise curve, rather than the simple one in figure 7, a curve that can be constructed by measuring the ratio of radiocarbon to normal carbon in many samples of known age.

Suess had developed improvements in Libby's laboratory techniques and was thus able to make more precise measurements or work with smaller samples. With these improved techniques Suess noticed what seemed to be a large deviation in the upper end of the calibration curve. Some modern material—the outer layers of trees that had just been cut—tested several hundred years old by his radiocarbon dating technique. Apparently, the most recent part of the calibration curve, instead of looking like the upper end of the curve in figure 7, was more like that in figure 8.[19] Had these grand forces of nature, just described, gone awry? Had the Milky Way suddenly decreased its production of cosmic rays or was the sun permitting fewer cosmic rays to enter the solar system?

The answer was informative. No, there had not been a decrease in radiocarbon in the air; instead there was a growing surplus of ordinary carbon, put into the air by the burning of fossil fuel. This surplus had diluted the radioactive carbon and

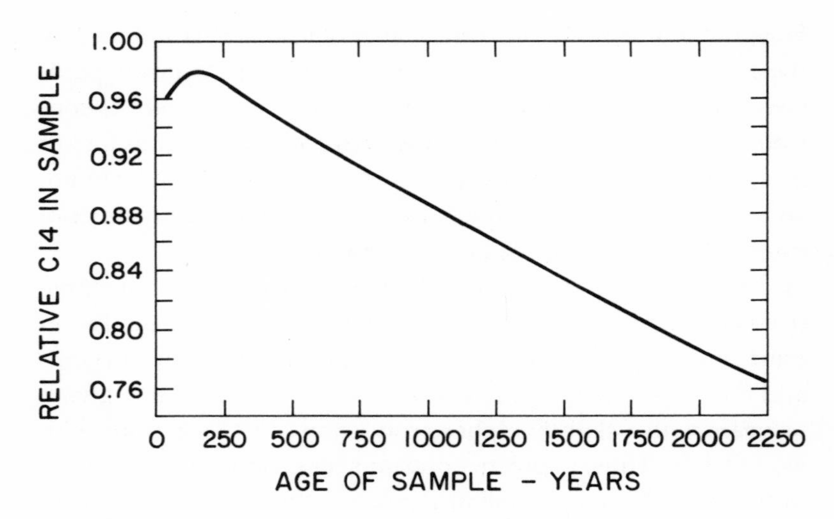

FIGURE 8. A calibration curve for radiocarbon dating suggested by Suess's measurements on modern material. In order to emphasize the decrease in recent years, this graph covers a smaller range of ages than does the line in figure 7.

reduced the radiocarbon ratio in things that had grown recently. People, Suess realized, had changed an important and measurable feature of the entire atmosphere of the globe.

In writing about his discovery Suess and a colleague, Roger Revelle, both distinguished scientists, remarked that "human beings are now carrying out a large scale geophysical experiment of a kind that could not have happened in the past nor be reproduced in the future. Within a few centuries we are returning to the atmosphere and oceans the concentrated organic carbon stored in sedimentary rocks over hundreds of millions of years. This experiment, if adequately documented, may yield a far-reaching insight into the processes determining weather and climate."[20]

This statement has been quoted many times since it was published in 1957, but with a clear evolution in tone. The original

quote is detached and scientific; we can make good use of this accidental experiment, it seems to say. Later versions take on an increasingly apprehensive note; we are experimenting with the whole earth without knowing how it will come out, they imply.

All of us can feel sympathy for future archaeologists who will find the usefulness of an important tool diminished, but we can content ourselves with the thought that they are a talented bunch of people who will surely figure out other ways to date their samples.[21] The other implication of Suess's finding is much more ominous: people are now so powerful that they can change the composition of the entire atmosphere. This discussion constitutes an appropriate introduction to the third of the three famous problems of the atmosphere: climate heating.

The change in atmospheric composition that Suess observed is still in progress, and changing faster than ever. The scientific community has not only attempted to insure that the change is "adequately documented"; it has tried to estimate what it means for the future of humanity. As a result, there exists a growing consensus among scientists that the changes under way in the composition of the earth's atmosphere will cause a rapid heating of the earth's surface and that at some time in the next century the climate will be hotter than ever before in human history.

This heating phenomenon is usually called the "greenhouse effect" or the "carbon dioxide" problem. But the physics of the situation on the earth is not parallel to the operation of a greenhouse, and carbon dioxide is not the only gas that can produce a climate change. For the purposes of this work, I will call the process "infrared trapping" and the effect "climate heating."

There is no controversy over the fact that infrared-trapping gases are being found in increasing concentrations in the earth's atmosphere. What may turn out to be the most important geophysical measurement of the twentieth century, shown in figure 9, is an extension of Suess's discovery.[22] This diagram shows carbon dioxide concentrations in the atmosphere, measured at the Mauna Loa Observatory in Hawaii, since 1958. This observatory sits high on the lava slopes of one of Hawaii's volcanoes, far from vegetation, towns, industries, and most people. The air samples are drawn from high above the site to avoid contamination by the few scientists and technicians who work there. In both the isolation of the sampling site and the care with which the samples are measured, the results are as representative of the amount of carbon dioxide in the clean air of the Northern Hemisphere as is practically possible.

The curve in figure 9 shows a large annual oscillation, which arises from the fact that plants absorb carbon dioxide from the air in the spring as they grow and return it to the air in the fall when they decay. But the most striking feature of this curve is not the yearly fluctuation but the steady increase over the entire period

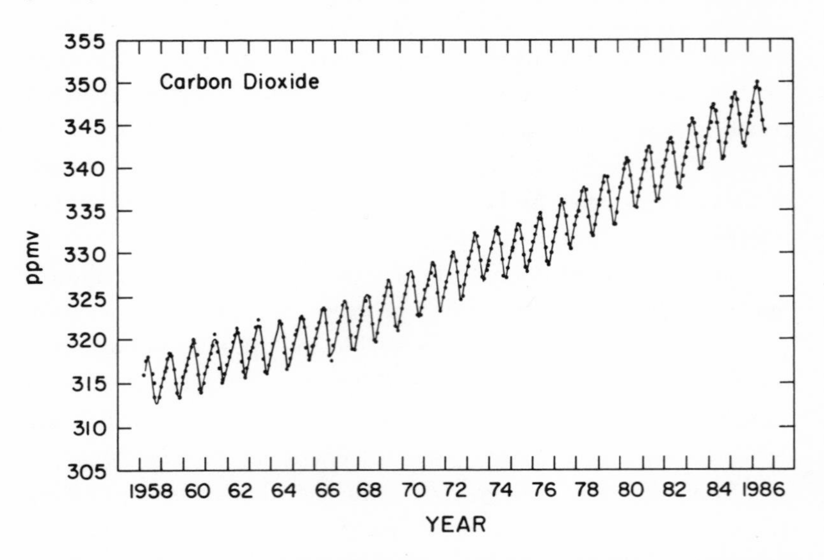

FIGURE 9. The concentration of carbon dioxide in the air as measured at the Mauna Loa Observatory from 1958 to 1986. Units of concentration are parts per million by volume.

of measurement to a current value rising above 350 parts per million. These measurements imply that there are now about 750 billion tons of carbon in the air in the form of carbon dioxide and that this amount is increasing each year.

These detailed measurements go back only to 1958, but scientists have found, by returning to the polar ice, a way of estimating the amount of carbon dioxide in the air at earlier times. Just as the lead from modern gasoline and the sulfate associated with acid rain are preserved in the ice, samples of air are trapped in the arctic and antarctic snows and are available for analysis. The polar ice takes years to seal off the bubbles of air completely, so the samples cannot preserve the month-to-month variation seen at Mauna Loa. But they do show the average value of carbon dioxide concentration during earlier centuries. Figure 10 comes from just such measurements.[23] From it we learn that the rapid increase in carbon dioxide started around two hundred years ago and has accelerated in recent times.

Suess's measurements created some surprise, but the idea

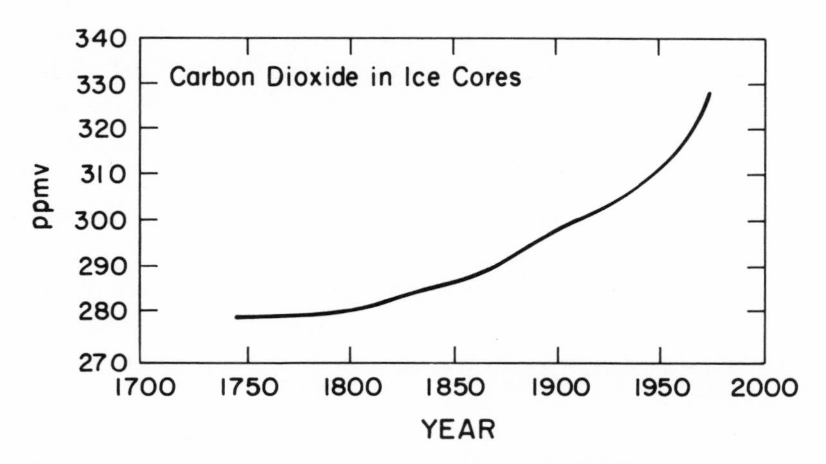

FIGURE 10. The concentration of carbon dioxide found in air trapped at different depths in polar ice. The concentration is in parts per million by volume.

that the burning of fossil fuels will cause carbon dioxide to accumulate in the air and that this accumulation will eventually heat the earth's surface is an old one. It was proposed at the end of the last century by Svante Arrhenius in Sweden and has been studied with increasing interest since that time.[24] For the most part, the progress in these studies can be characterized as an increasingly sophisticated appreciation of the nature of the climate system, its internal workings, and its interactions with the ocean, land, ice, and small changes in the earth's orbit around the sun. But one major change in the description of the problem occurred recently. Experts on atmospheric radiation realized that some gases other than carbon dioxide could also trap infrared radiation, and experts in the chemistry of the air showed that the concentrations of some of these "other gases" were increasing in the atmosphere. With this new realization, it became clear that carbon dioxide was just the first on a growing list of substances that could be viewed with apprehension.

Foremost on this list of other infrared-trapping substances is methane. Modern measurements show that the atmospheric concentration of this gas is increasing about twice as fast as that of carbon dioxide. Once again, the polar ice lets scientists determine when this increase started. Figure 11 reveals that the upward swing started at about the same time as the carbon dioxide increase and, similarly, has accelerated in recent times.[25]

The next two items on the list are CFC-12 and CFC-11—the same gases that stand accused of depleting stratospheric ozone. The concentrations of these gases are increasing very rapidly, as shown in chapter 3. Since these gases were first synthesized in the laboratory in this century, their increase in the air is a recent phenomenon.

The list gets longer each year as chemists develop new products that are then manufactured in large quantities. As yet, most of the newer substances have not appeared in the air in trou-

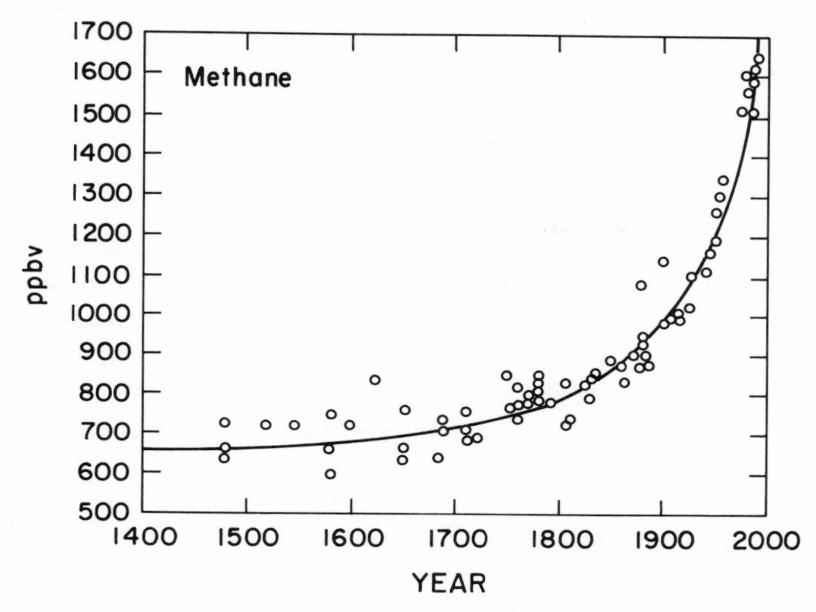

FIGURE 11. The concentration of methane in the atmosphere at various times in the past, as deduced from measurements of air trapped in ice cores (1480–1950) and from direct measurements of air samples (after 1950). The concentration is plotted in parts per billion by volume.

blesome amounts. But two other gases, which are not newly synthesized chemicals, are high on the list of infrared-trapping gases worth watching. These are nitrous oxide and ozone in the lower atmosphere. The concentration of nitrous oxide in the air is increasing at a measurable rate, although more slowly than carbon dioxide. The concentration of ozone in the lower atmosphere is thought to be increasing, but because its lifetime in the lower air is short and the variations in its concentration from place to place are large, it is difficult to say for sure that an overall increase is under way. But for all the gases high on the list, except ozone, measurements show clearly that the atmospheric concentrations are growing, and for each of these the increase is closely correlated with the rise of industrial societies and the growth of the earth's population.

INFRARED TRAPPING

Viewed from afar, the earth must radiate away the same amount of energy that it absorbs from the sun, and this balance must continue no matter what changes we make in the air. But the details of how the earth arranges to radiate that heat away are strongly influenced by the composition of the atmosphere.

Although it might seem unlikely that gases making up a few hundredths of one percent of the atmosphere could affect the climate, it is indisputable that infrared trapping and climate heating operate currently in the atmosphere of the earth and have done so throughout the history of the earth. If all infrared-trapping gases could be removed from the air and nothing else changed, the surface of the earth would have an average temperature of about −18° C.

When the surface of the earth sends heat or infrared radiation upward, the heat-trapping gases absorb much of it before it can escape to space. These gases are themselves warmed by the radiation they have absorbed, and they in turn radiate this extra energy away in all directions. Some of it therefore returns to and warms the earth's surface. The radiation that is able to leave the earth mostly originates high in the atmosphere where the temperature is cold, −18° C, thereby maintaining the overall balance with arriving sunlight. The average temperature of the earth's surface is about +15° C; the infrared trapping occurring today warms the earth's surface by an average of 33° C, from well below freezing to well above. Were this not the case, the earth would be permanently covered with ice—an inhospitable place for human existence, or indeed for human evolution in the first place. The "leaky barrel" analogy from chapter 3 is an appropriate way of looking at the problem of climate heating. The process of adding infrared-trapping gases to the air and thereby raising the temperature is analogous to plugging up a few of the holes in the barrel, thereby forcing a higher water level in order to restore the total leakage to a value that is in equilibrium with the input.

The fact that the historical concentrations of infrared-trapping gases heat the earth's surface by a large amount immediately raises a question: Will an increase in these concentrations raise the temperature even more? The answer is almost certainly yes; the only question is by how much.

Scientists again resort to complex numerical models as the best method of calculating how much the climate will change as a result of past emissions and how much more it will change if we continue to emit heat-trapping gases into the atmosphere.[26] Even the simplest model of the climate—one based on a global average atmosphere, measured values of incoming solar energy, the optical properties of the gases in the atmosphere, and the rate of temperature decrease with altitude—gives a fairly good value for the current amount of climate heating. But difficulties become apparent if one uses such a simple method to compute how much the surface temperature will rise as the amount of carbon dioxide and other infrared-trapping gases in the air increases. For example, the results of the simple calculation are very sensitive to the amount of water vapor in the air, because water vapor is also a powerful absorber of infrared radiation, more effective in fact than carbon dioxide.[27] As the surface air heats as the result of the added trace gases, it is reasonable to suppose that the amount of water vapor in the air will increase, evaporated from the oceans by the higher temperature. (This process takes place every year; on the average, there is more moisture in the air in the summer than in the winter.) And if there is more water vapor, the surface will be made warmer still. To include this secondary process in the calculation requires a much more complex procedure, one that goes into more detail than a one-step calculation averaged over the whole earth.

The additional detail needed includes not only information about how much water vapor is added to the air but also where it is. Conditions differ from one place on earth to another. Radiation reflected or absorbed and radiation emitted into space strongly depend on local circumstances. The amount, kind, and height of clouds, the color of the ground, the temperature of the ground or air, the presence of snow cover—all influence the local streams of radiation. And although the concentration of carbon dioxide in the air is very nearly the same everywhere, the amount of water vapor is highly variable. Furthermore, water vapor, heat energy, momentum, dust, and other materials or conditions of the atmosphere can be moved from one place to another by winds and from one height in the air to another by convection, making the complete calculation of how the atmosphere behaves complex indeed. The calculations required to capture these complexities involve simulating not only all the important processes at work in the air at one moment, but also the changes in these processes every few minutes. Such climate models require teams of scientists and computer programmers, working for years, for their design and construction, and millions of dollars' worth of computer time for their testing and operation. These costs limit the number of scientific groups that can create climate models; as a result there are fewer than a dozen reasonably independent models in use in the world that can produce heating projections.

Modelers look at the climate in many different ways, but they all employ one standard calculation so that their results can be compared with those from other groups. They first calculate weather conditions for a simulated time of ten years or more, long enough so that average temperatures, rainfalls, and other quantities can be obtained. The first calculation is made with a standard atmosphere—one with the current amount of carbon dioxide or with some nominal earlier amount, say, 300 parts per million, volume (ppmv). Then a second calculation is made with twice as much carbon dioxide, 600 ppmv. The two results can be compared and the impact of the extra carbon dioxide estimated. Each modeling group may do many other calculations, depending on its particular interests, but the standard calculation is the one that is gathered by reviewers and international committees wishing to see if a consensus is developing on the probable impact of an increase in atmospheric carbon dioxide.

The authors of the most recent review were careful, as were their predecessors, to examine the differences between the results of different models and different research groups.[28] Some models show larger amounts of heating than others; some show increases in rainfall at places where other models show no change; some models show larger increases in temperature in the Arctic than others. But at the core there is firm agreement: All the models show an increase in climate heating. All the models show the heating to be greater at high latitudes than near the equator. And all models show that while the surface temperature of the earth will rise, the stratosphere will get much colder. This survey concludes that the average amount of global heating of the earth's surface that would result from a doubling of the amount of atmospheric carbon dioxide lies in the range of 1.5 to 5.5° C, with an estimated two-thirds probability that the correct answer is between 2.5 and 4.5° C.[29]

This custom of comparing and reporting on the results of the standard experiment as performed by each model is useful and convenient, but it can lead to misunderstandings of two different kinds. The first arises from the fact that the standard experiment focuses the reader's attention (and sometimes that of the authors) on carbon dioxide, with the result that the effects of the other heat-trapping gases may be forgotten. The voluminous and careful study by a committee of the National Academy of Sciences in 1983 fell into just such a trap.[30] Although the report of this committee included a short section on the effects of non-CO_2 gases, these other gases were forgotten when the summary and recommendations were written, and a much more relaxed time scale for the onset of changes was implied than could be justified by the data in the report.

The second misunderstanding that arises from the standard experiment is the impression it creates that the amount of carbon dioxide will double, the climate will heat by 3 degrees, and that will be the end of it. In fact there is no known process that will cause carbon dioxide concentrations to level off at 600 ppmv and the climate heating to assume a stable value as estimated by the models. All projections of future emissions of heat-trapping gases lead to the conclusion that the climate will continue to heat as concentrations of these gases continue to rise. The policy issue is therefore not how will we live with a climate that is 3 degrees warmer, but how can we adapt to a climate that is continuously changing.

To avoid this kind of mistaken impression, recent review groups have begun to describe their conclusions in terms of rates of change, rather than as fixed amounts of heating. The most recent Villach study reports that the earth's average temperature is likely to increase at about 0.3° C per decade in the coming years, accelerating later if rates of emission of heat-trapping gases continue to rise.[31]

THE IMPORTANCE OF CLIMATE HEATING

None of us are accustomed to thinking in terms of global average temperatures. We do know that we experience large temperature changes between summer and winter and that we can travel to the tropics or the Arctic without doing ourselves harm. So it is reasonable to wonder what an increase of 0.3° C per decade means. And even if scientists are able to convince everyone that such a shift represents a major climate change, it is still legitimate to ask whether humans—a notoriously adaptable species—cannot arrange to adapt to any changes these shifts in climate will bring. It could well be that the amount of heating foreseen will simply mean that farmers will choose different varieties of crops to grow, summers will be a bit longer and winters a bit shorter, and dealers in heating fuels will need to diversify into air conditioner maintenance. However, it might instead mean that the heating could produce major disruptions in our modern civilization.

One way to consider the question of the importance of the coming climate heating is to look at past climate changes that bear some similarity to the projected ones and see whether people have generally been successful in coping. We can ask how much and in what way the climate changed, and how these changes affected the people who lived through them.

Good records are available for both the climate and human activities at various places around the world for the past century or two. But there are distinct advantages to looking back ten thousand years. This length of time more than covers the whole history of organized, record-keeping civilizations, and it includes at its earliest point the retreat of the most recent glaciation, the largest long-term climate change that we can reconstruct in some detail. Between ten thousand years ago and today there have been other changes in climate, some of them apparently global in extent, and their impacts can be studied, with increasing reliability as we approach modern times, when the invention of the thermometer, the barometer, and the routine measurement vastly improved the picture of the climate and its changes. For most of the ten thousand years, however, climatologists are dependent on what they can deduce from indirect evidence to sketch out a picture of the climate at some particular time.

The detailed records of the last hundred or more years serve as a lesson on the kinds of problems that may be encountered in piecing together a climate history for a longer period. Figure 12 shows the average temperatures measured at a station near Saint Louis, Missouri, during a 140-year period up to the early 1980s. The temperature there increased, somewhat erratically, until about 1931, dropped from 1931 to 1977, and then began a rapid rise.[32] Figure 13 shows a similar set of measurements from Denver, Colorado—a slightly shorter set, starting in 1873, with the same rise to 1931 as in Saint Louis (see n. 32). But in Denver there was a second, higher peak in 1955, and the rapid rise in the 1980s is not apparent. A record from farther east, a station in Colum-

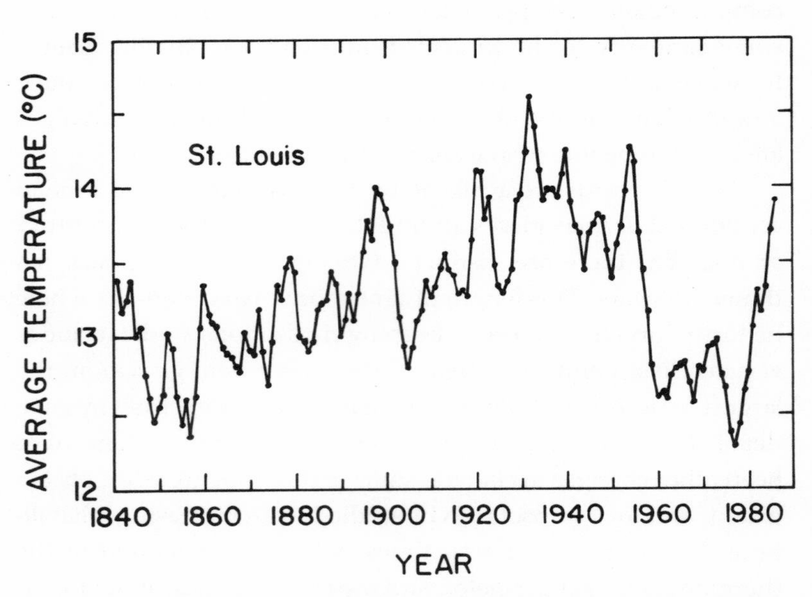

FIGURE 12. The temperature measured near Saint Louis, Missouri, for the period 1840 to 1987. The single-year measurements vary by as much as 2 degrees, so the five-year running averages are plotted here to eliminate some of the larger year-to-year variation and to emphasize longer-term trends.

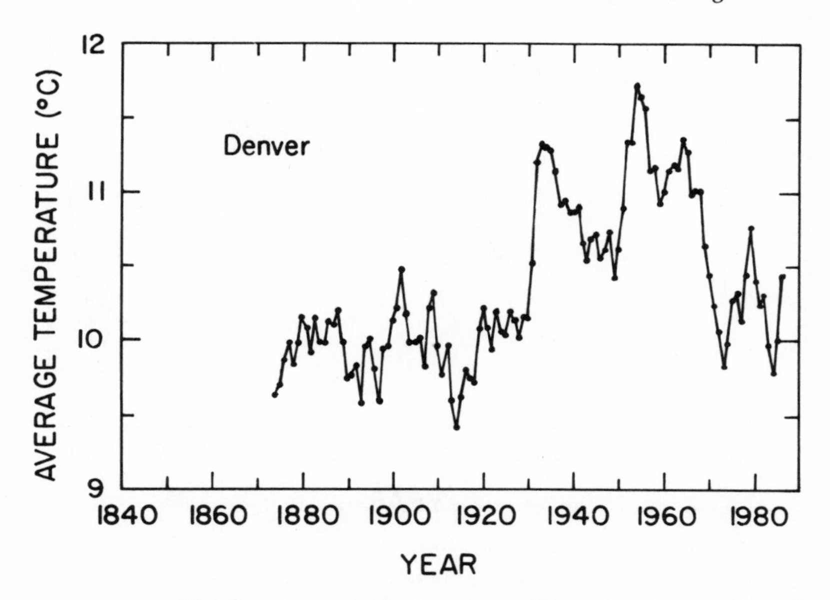

FIGURE 13. The five-year running average of temperatures measured in Denver, Colorado, about 1,275 kilometers west of Saint Louis.

bus, Ohio (fig. 14), shows the same peaks as Saint Louis and Denver and also the recent rapid rise.

The similarity in these three records, based on measurements at locations more than 2,000 kilometers apart, suggests that such climate variations as the 1930 warm peak may be widespread. But these changes were not global; they do not even characterize the changes everywhere in the Northern Hemisphere. Figure 15 shows the measurements from Palma, on the island of Mallorca in the western Mediterranean. The pattern presented by these observations is almost the opposite of the one found at Saint Louis.[33]

To find out what was happening to the global climate during these years, it is necessary to gather information from hundreds or thousands of measuring stations, distributed widely over the

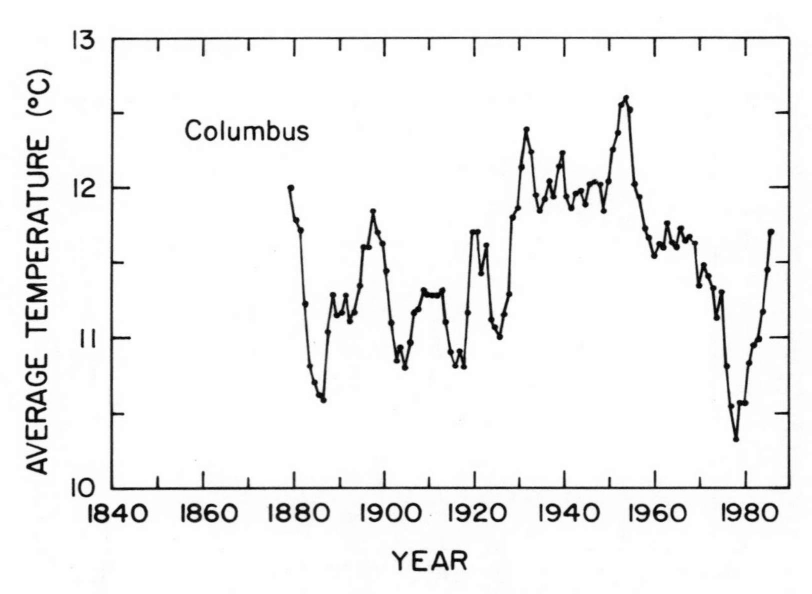

FIGURE 14. The five-year average temperatures measured in Columbus, Ohio, about 845 kilometers northeast of Saint Louis.

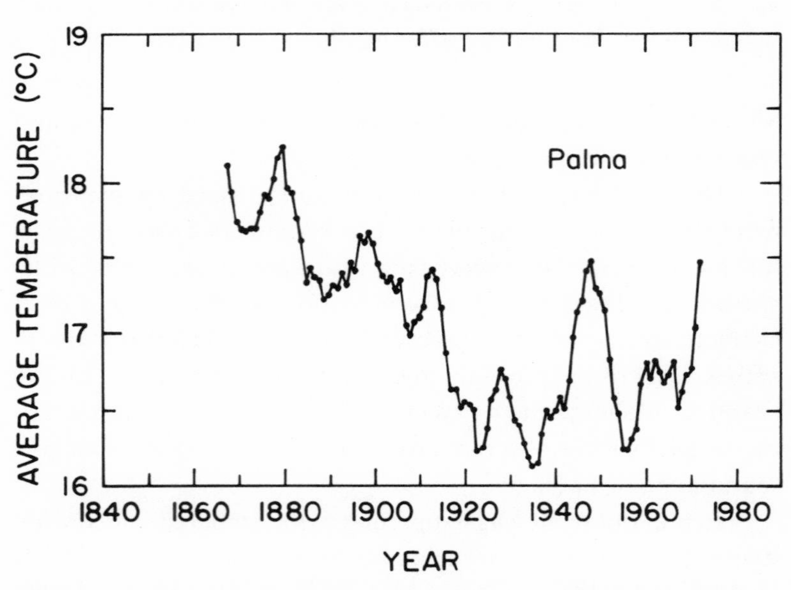

FIGURE 15. The five-year running average temperatures measured in Palma on the island of Mallorca in the western Mediterranean Sea.

earth, and average them together. The process is tedious, since the location of each station must be considered to make sure it is not unduly influenced by changes in local human activities—a factory built near a measuring station can strongly affect the readings. Data from closely placed stations must be compared to help find recording errors, and much care must be exercised to catch occasions when a station was moved or a thermometer became defective. Then the measurements can be averaged for an estimate of the climate. A major difficulty in this process is that most of the earth's surface is ocean, and there are very few stations at sea that have been keeping records long enough to be added to the land-based records. One technique for overcoming this difficulty is to take measurements made from ships crossing the ocean—either measurements of the temperature of the air near the ocean surface or measurements of the temperature of the surface water itself—and include them in the average at appropriate locations. Figure 16 shows a Northern Hemisphere average developed in this way.[34] The changes from one decade to the next are much smaller than those at a single station, such as Palma or Saint Louis. A vestige remains of the large peak near 1930 seen in the records from the central United States, but the outstanding feature of the hemispheric average is the clear difference between the years near the beginning of the record and those near the end. The average for the Southern Hemisphere (fig. 17) shows the same increase over the last 120 years (for source, see n. 34). The earth warmed on the average a total of about 0.5° C during that period, but with much larger fluctuations in particular regions.

Thermometer records from widely placed locations do not go back much beyond 150 years, so in order to examine the changes that may have occurred in the climate during the last thousand years, other sorts of evidence must be used. For the literate, densely populated portions of the earth, written records can be used to deduce some aspects of the annual weather. The freezing

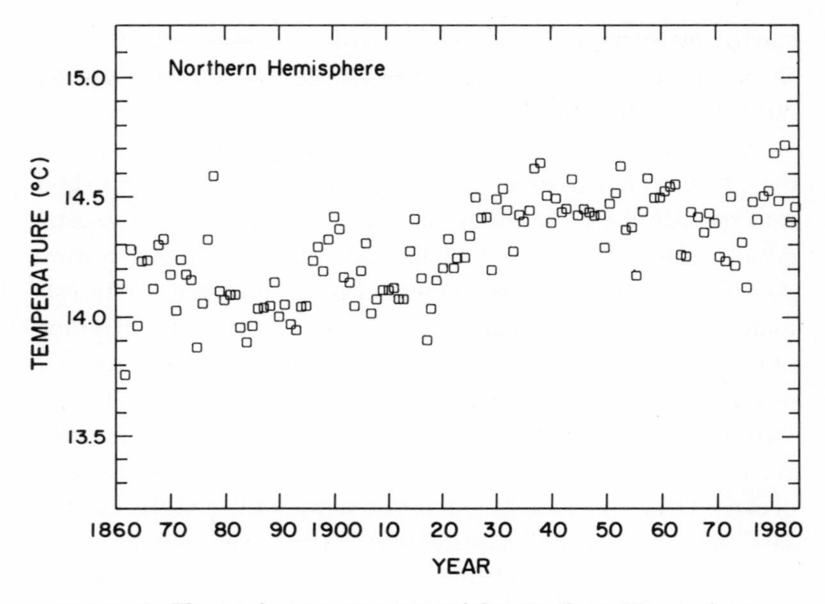

FIGURE 16. The surface temperature of the Northern Hemisphere since 1861, as deduced from measurements over land and ocean areas and of the temperature of the water at the ocean surface.

or thawing of certain harbors and canals are clearly temperature-related events, and helpful hints, though usually unaccompanied by numbers, can be gleaned from diaries or letters claiming that "this was the coldest winter in the memory of the oldest inhabitant." In addition to these documentary sources of temperature information, measurements are available of the width of tree rings, which are responsive to temperature in some locations. The experience gained from examining the American and Mediterranean thermometer records cautions us that information from a single location should not be interpreted as a hemispheric or global change and that changes at a fixed location are likely to be considerably larger than the global change.

Figure 18 shows two forms of deduced temperature for the

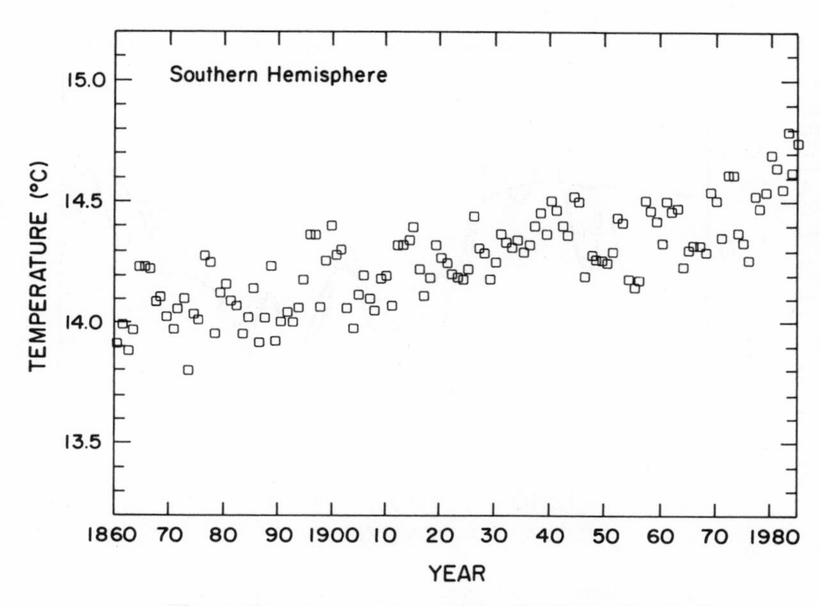

FIGURE 17. The surface temperature of the Southern Hemisphere since 1961, as deduced from measurements over land and ocean areas and of the temperature of the water at the ocean surface.

last millennium. These records come entirely from the Northern Hemisphere, and they pertain mostly to Europe and North America, but they have so much in common that they must represent a widespread climate pattern even if not a clear global change.[35] The most prominent features in the period since A.D. 989 are the warm interval around 1200, and the two cold eras around 1400 and in the seventeenth century. The first is called the Medieval Warm Epoch and the two cold periods make up the Little Ice Age.

These were large changes in climate. A longer and harsher winter season and longer periods when harbors were iced over influenced in some degree the activities of people. Some marginal societies were destroyed. For example, during the Warm

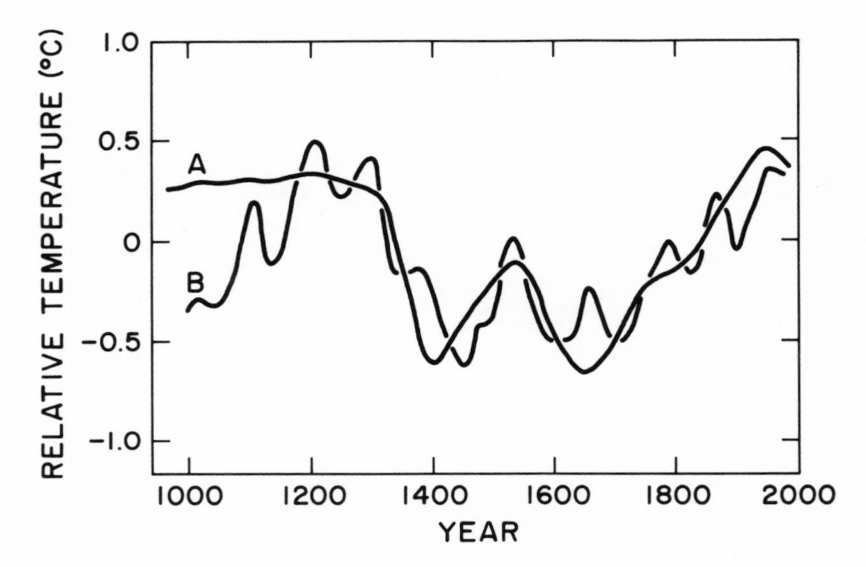

FIGURE 18. Estimates of temperature for Northern Hemisphere locations during the last one thousand years. Curve A was derived from various records of the severity of the winter in Europe. Curve B was derived from the width of tree rings from the West Coast of the United States.

Epoch, Norse explorers established colonies in Greenland. These groups grazed cattle on grass growing along the river bottoms and survived for hundreds of years, with only occasional contact with the Norse homeland. But when the Little Ice Age came, the grass, the cattle, and the settlements disappeared.[36] In North America there is evidence that the cooling after the Warm Epoch occurred quite suddenly and that within a century villages that had existed in the northern high plains disappeared.

But climatic events deduced from archaeological reconstructions are rare. Historians who have hoped to explain some of the shifts that occurred in the progress or success of particular civilizations as responses to climate change have found such impacts elusive. Everyone agrees that years of particularly extreme weather can have an immediate and measurable influence on people. A major crop failure, as in 1740 in parts of Europe, after years of good growing seasons may catch people unprepared and result in great hardship. But from a greater distance, it is not obvious that the bad year had any lasting effects. Some historians claim to see the effect of climate in events as great as the decline of the Late Bronze Age civilization of Mycenae three thousand years ago and the much more recent transatlantic movement of European populations; other scholars point out that other trends were also under way that could equally well have caused the cultural or economic shifts. One historian even writes: "In the long term the human consequences of climate seem to be slight, perhaps negligible, and certainly difficult to detect."[37]

These differences in opinion stem in part from a difference in what is regarded as a "human consequence." It is clear that a climate change can force people to do things differently than they did them before. If they adapt appropriately, historians will find no big changes in the incidence of hunger or starvation, no increased mortality, and no loss of economic or military advantage relative to other peoples, and may therefore conclude that the climate change had little effect. Another person studying the same episode may note that the people made significant modifications in their schedule for planting, the crops they grew, the trades they negotiated, and the hours they worked in order to live with the new climate, and deem that the climate change had large human consequences. At worst, a sudden climate change catches people off guard and produces suffering; even a slow change can push vulnerable societies over the edge. At best, people adapt readily to slow changes by modifying their activities.

Beyond one thousand years ago, the documentary evidence available virtually vanishes. The climatologist becomes almost completely dependent on what can be deduced from ice, trees, ancient seashores, and other data that can be interpreted in terms of climate. The Greenland ice can again be helpful. When snow forms and falls on the ice cap, it preserves not only impurities, such as sulfate and lead, and air composition, such as the proportions of carbon dioxide and methane, but also something about its own composition. Water, H_2O, contains a small fraction of molecules that have a heavier form of oxygen, oxygen 18, and an even smaller fraction containing deuterium, a heavy form of hydrogen. These molecules, being heavier, move a bit more slowly and are incorporated into snowflakes at a different rate from that of the common water molecules, the difference depending on the temperature. Thus, a measurement of the fraction of the melted snow that contains oxygen 18 or deuterium from various depths in the Greenland ice can indicate temperatures long ago.

The result of such ice-core investigations gives us an estimate of temperature change (fig. 19).[38] The most notable aspect of figure 19 is the warming beginning about fourteen thousand years ago, an event quite clearly associated with the retreat of the last ice age. The retreat was apparently nearly simultaneous in both north and south polar regions and represents a change of greater magnitude than any that has occurred since. It took place over several thousand years and is still affecting the earth. In locations that had carried great depths of ice, the earth was depressed by the weight. As the ice melted, the weight was removed, and the depressed land began to rebound or rise to a new level. This slow process is still under way. In Scandinavia, for example, an apparent slow drop in sea level today is actually the continued rebound of the land from the last glaciation. But more important for this discussion is the fact that the climate change associated with the retreat of the ice was large enough to have major consequences for a basic aspect of our environment. The character of this impact emerges from the study of pollen grains deposited in lakes and bogs during the time of the retreat.

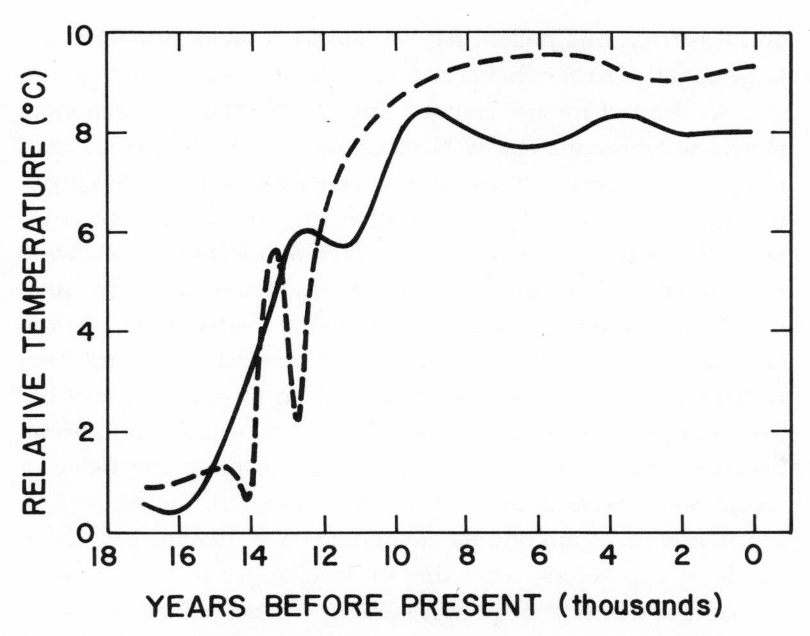

FIGURE 19. Two estimates of the temperature for the last seventeen thousand years. These curves were derived from measurements of the amount of oxygen 18 in the ice of Greenland (dashed line) and Antarctica (solid line).

Pollen grains are very durable, and the plants that produced them can frequently be identified by a careful examination of the grains under a microscope. The date at which they were deposited in the lake or bog can also be established in many cases from radiocarbon analysis of the organic material at various depths. To some extent, this pollen can be used to bolster our information about ancient climates. If the dated and identified pollen comes from a grass or shrub that cannot tolerate winter frost, we can be sure that the winters were mild at that time. If it comes from plants requiring a great deal of water, we learn something about the rainfall. The information gleaned from pollen must be used with care, however, since the reactions of some plants—trees in particular—to a changed climate may be delayed by centuries, and it is from tree pollen that we deduce a major impact of the large climate change that occurred fourteen thousand years ago.

As the last ice age receded and the Northern Hemisphere slowly warmed, the face of North America was completely rearranged. Tree species that grew in what is now Ohio and Michigan migrated far up into Canada to find a climate they could thrive in, and other kinds of trees from the deep South moved into Ohio and Michigan. Even after the ice had retreated to northern Canada, changes continued in what is now the United States. Grassland moved east into Indiana for a thousand years and then returned west, to be replaced by trees in locations east of the Mississippi River. In what is now Denmark, some plant species became extinct when they were unable to jump the Baltic or North Sea and continue their northward march.[39]

These mass migrations, sometimes over distances of thousands of kilometers, were driven by changes in average temperature of about 10 degrees occurring over more than two thousand years, or a rate of temperature change of 0.05° per decade or less. During the Little Ice Age, and for the most recent one hundred years, the hemispheric or global temperatures changed at rates of up to 0.05° per decade, but these changes did not continue for much more than a century and thus did not result in total changes of more than a degree. These numbers give us a way of estimating whether or not the climate heating that will follow increases in the infrared-trapping gases will be serious.

At the present rate of increase, carbon dioxide will reach a concentration of 600 ppmv sometime in the third quarter of the next century. The other infrared-trapping gases will approximately double the effect of carbon dioxide. From these numbers the rate of heating can be estimated to range somewhere between 0.2° to 0.5° per decade, or from four to ten times as fast as these earlier changes.

This rate of change is much faster than any past climate event involving human civilizations, so there is no direct information on the kinds of disruptions such changes could cause to human society. But the migration of forests can provide some clues as to what kinds of changes to expect. It seems unlikely that most tree species will be able to migrate fast enough to keep up with accelerated changes. Mature trees can withstand large changes in conditions, but by the time a young tree has grown large enough to produce seeds, the climate would no longer be suitable for a new tree to take root, and so many forest species will disappear. The damage to trees will be even greater in regions in which plants have already been weakened by acid rain in the soil and the impact of ozone and other pollution on the leaves or needles.

Changes in sea level will also be serious. During the peak of the last glaciation, much water was stored in the form of ice on Greenland, northern Canada, Antarctica, and high mountain ranges. As a result, sea level was far below what it is today. Much of that ice has now melted, but some remains; continued warming will bring continued melting and a continued rise in sea level. Estimates of the rate of rise, like the estimates of temperature increase, show an acceleration to values of ten times that experienced in the last century, with sea level at the end of the next century projected to be a meter higher than it is today. In the United States the first regions to be affected will be those already experiencing difficulty, such as eroding East Coast beaches and large regions in southern Louisiana. Near the mouth of the Mississippi River, in Louisiana, the ground is sinking slowly. Earlier, frequent deposits of fresh sediments from the flooding river compensated for the sinking. But the Corps of Engineers' success in managing the river, preventing floods, and providing continuous deep channels for ships has cut off this supply of new earth and brought to light the gradual subsidence of the area. In a region such as this, an increase in sea level of even a few centimeters can make it impossible within reasonable costs to save large areas of land from permanent inundation.

All estuaries will need to move inland as seas deepen,

forcing people to choose between allowing the migration and giving up what is now farm, forest, and city, or building dikes to prevent the movement and giving up the high seafood productivity that is characteristic of estuaries. Even if free movement is allowed, some estuaries will be unable to migrate inland because the terrain is too steep. Recent studies estimate that the United States will lose half its present coastal wetlands in the coming century.[40]

The list of other possible impacts is a long one. Crops adapted to a particular temperature range will need to be replaced with new ones, while the old ones will need to be moved nearer to the pole. Warmer winters will interfere with mountain snowpack used for irrigation and will allow some pests that are now controlled by winter freezes to survive. Growth will be slowed in tree species that require extended periods of winter cold in order to trigger their springtime resurgence. With a different temperature distribution over the earth, there will also be significant changes in the rainfall pattern. Even areas that get the same amount of rain will find that, with higher temperatures, evaporation of water at the earth's surface will increase and more rain will be needed just to maintain the same soil moisture during the growing season as before.

Many more impacts of climate change have been the subject of calculation or speculation. But the first question that needs to be answered is just how much confidence should be placed in these projections of a rapidly changing climate.

5. But Is It True?

The earth, with its atmosphere and oceans and living things, is such a complicated system that any analysis of one interaction—for example, how the composition of the atmosphere relates to the temperature of the earth's surface, or why a change in the amount of ozone in the stratosphere might affect the health of ocean plankton—is both hard to understand and easy to doubt. So many other features of the system are excluded from examination that it becomes reasonable to suspect that the discussion—especially if it concludes that we must take action—is not an accurate forecast of the future but rather an artifact determined by the study's particular selection of which parts of the overall system to consider. Until this suspicion is overcome, no amount of warning about the need for early action is likely to be heeded. The purpose of this chapter is to convince the reader that, despite these difficulties, there are compelling reasons today for believing that major changes of human origin are about to unfold and for concluding that the warnings must be taken seriously.

The magnitude of the human influence strongly supports such a belief. One would be amply justified in doubting that stamping his foot on the ground could trigger an earthquake. But a scientist's speculation that one earthquake could trigger another quake in a nearby fault would deserve serious study. The problems in the atmosphere resemble the latter situation: the collective activities of people now rival or exceed conditions that have governed the climate and the interaction of the atmosphere with living things for ages. We have affected plant and animal life at many locations as we have tripled the amount of sulfate circling the Northern Hemisphere. We have disrupted the energy flows in the air more strongly than volcanoes or sunspots do. And our impact on the ozone layer has been so large and so sudden that it has caught us by surprise. We have forced all of these changes by means of substances that our vigorous industrial and agricultural activities release into the air. If we still have doubts that we are that powerful, we have only to remember that next year our releases will be even larger and our effect on the atmosphere even stronger.

The second reason for believing the impacts foreseen by scientists is that they are already happening. We can see that sulfur is killing lakes, and we can measure the ozone drops over Antarctica. The most pervasive impact of all, climate heating, is not an esoteric possibility imagined by climatologists: life on earth depends on its existence, as described in chapter 4, and the additional heating foreseen for the coming decades is therefore a logical extension of well-understood phenomena. Furthermore, in studying climate heating, we are not confined to the one case of our earth; we can observe two other planets, similar in size and situation to ours, one of which has much higher concentrations of infrared-trapping gases in its atmosphere than we do and one that has much less. Mars has almost no infrared trapping, and its surface temperature is $-53°$ C, about what one would expect from its distance from the sun. Venus has many times the amount of carbon dioxide in its atmosphere that earth does, and its surface temperature is $+427°$ C, far hotter than can be explained by its proximity to the sun. The earth is intermediate; it has some infrared trapping in its atmosphere, and its surface temperature is between that of the other two. The evidence, then, that climate heating is a genuine problem assumes more strength: climate heating works here (33° C on earth today) and works over a wide range of conditions (3° C heating on Mars and 467° C heating on Venus).

These facts are impressive but insufficient. The earth differs from its neighboring planets in being mostly covered by water—water that stores heat, water that provides clouds, water that makes snow and ice. Even though Mars and Venus convince us that infrared trapping is real, the problem of learning just how much the climate will heat on earth involves processes that cannot be observed elsewhere and must be calculated based on our understanding of our climate system. In other words, we are again forced to use numerical simulations or models, much as we were for acid rain and stratospheric ozone depletion. As I reported in chapter 4, the climate models foresee a heating that will raise the surface temperature as much in the next few decades as was seen during the century-long recovery from the Little Ice Age, and ten times as fast as it did during the long retreat of the last glaciation.

Not surprisingly, when a modeler tells members of the public that activities that they enjoy—such as staying warm, traveling, and eating—are producing changes that they may not be able to adapt to, the standard reaction is one of considerable skepticism, and the model is asked to survive the most careful scrutiny.

In their struggle to produce usefully realistic models of climate heating, research scientists have added, one by one, complexities to their models as computers have grown more powerful and knowledge of the climate has grown more comprehensive. The use of a single average to represent conditions everywhere on earth was very early replaced by a more detailed calculation in which the atmosphere was represented by many points on the earth's surface and several points up through the depth of the atmosphere. The single calculation of a temperature gave way to hour-by-hour calculations of a changing weather pattern, and a long list of physical processes thought to be important has been included in the model calculations. Today, in these complex models, realistic winds develop, clouds form, precipitate, and disappear; air flows over mountains; moisture is exchanged with the ocean surface; soil moisture accumulates and evaporates; and snow and ice come and go with the seasons in high-latitude regions. Why then are there any doubts about the answers that such models give?

The reason is that deficiencies remain even in today's most sophisticated models. The most troublesome weakness lies in the way in which clouds are simulated. Clouds are simultaneously exceedingly complex and very important to the workings of the climate (and hence of climate models). The amount of solar radiation reflected back to space is critically dependent on the amount and kind of cloud cover, and the amount of infrared energy radiated from the top of a cloud to space depends on just how high in the atmosphere (and hence how cold) the cloud top is. But the formation and growth of a cloud or cloud system involves, among other things, subtle aspects of vertical motions of the air, water vapor saturation, the presence of nucleating agents, and the vertical temperature structure of the air. Once formed, clouds affect radiative balances depending on cloud thickness, the size of the water droplets, impurities in the water drops, and perhaps other quantities. Even the most advanced of today's climate models can create clouds only on the basis of a few rather simple ideas; for example, whenever the model finds the relative humidity higher than a certain value it makes a standard cloud at that location. Or it may be instructed to make a cloud where the air is unstable or when low-level winds are bringing fresh moisture to a particular place.

A second weakness arises from the way in which models represent the atmosphere. Computations for current climate models are carried out on the largest computers available, ones that perform a half-billion operations each second. Even so, to keep the calculations within the confines of available computing resources, the number of points used to describe the atmosphere must be kept lower than is desirable, resulting in a considerable blurring of the distinctions between mountains and plains, land and water, and ice-covered and open surfaces. For example, one of the elaborate climate models in use today employs about 1,900 points over the surface of the earth, each point representing about 270,000 square kilometers. One point must therefore represent the average conditions over all of Greece plus the Aegean Sea, or, in the United States, all of New Mexico from wooded mountains to white sand deserts.

Finally, oceans are poorly simulated in climate models. A model that realistically incorporates all important atmospheric phenomena, with fine spatial resolution, will still be inadequate to deal with all features of climate change over decades or centuries unless it can be coupled with a suitable model of the oceans. This vast expanse of water can absorb and give off water vapor and heat; it can move heat from one location to another; and it constitutes a large heat reservoir that can modify how rapidly the global temperature climbs. But compounding the difficulty of including these processes in the models is the fact that the oceans are not independent, external agents in the climate system; they are part of it. The ocean currents, which help determine the surface temperature of the sea, are driven by winds; deep ocean circulations are driven by such atmospheric phenomena as rain and cold air that change the density of sea water and thereby govern when water will sink to the deep layers and where it will reemerge to the surface. The air with its winds, the ocean and its circulation—each helps drive the other; each is dependent on the other, and the calculation involving the oceans must be sophisticated enough to allow for this two-way interaction.

The oceans also produce transient effects. The usual procedure, described earlier, for producing model calculations keeps the concentration of infrared-trapping gases fixed at the current or doubled value during the simulation of many years. But the amount of carbon dioxide and other infrared-trapping gases is changing continuously today and is likely to be changing twenty or fifty years from now. This would cause no great difficulty if it were not for the time lag in the heating effect that will be produced by the large heat reservoir of the ocean. The ocean-land distribution is very different in the Southern Hemisphere than in the Northern Hemisphere, and one can therefore expect that the lag produced by the ocean in the heating process will also be different in the two hemispheres. The Northern Hemisphere, with most of the earth's land, is likely to heat more rapidly than the Southern Hemisphere, with most of the world's oceans. This difference may produce changes in the general circulation of the atmosphere that are not revealed in the steady-state type of calculation described above.

Beyond these problems lies the possibility that the ocean circulation itself—driven by winds, rains, and evaporation—will undergo large changes as the atmospheric circulation evolves and thus alter such major features of the total climate system as the transport of heat away from the equator or the rate of heat transfer to the ocean. Scientists are not confident that the simulations of ocean phenomena that are incorporated in climate models are sufficiently detailed yet for the model to foresee such changes. Until the simulation of oceans and ocean-air interactions is improved, models will be useful only for foreseeing large-scale regular changes that proceed continuously away from today's climate, not the more violent and sudden changes that might occur if we push the system too far.

All these weaknesses together produce the most frequently cited limitation of climate models: as yet they show little ability to project the regional manifestations of the climate heating—how hot will Kansas be in 2030 and what will the average rainfall in the Sahel be in 2040? We know from studies of the small climate changes of the recent past that all areas do not warm and cool at the same rate. During the first half of this century, when the Northern Hemisphere was warming a half-degree (fig. 16), some mid-latitude locations warmed four times that much, while other locations cooled (figs. 12–14). While it seems likely that trace-gas-induced climate heating will result in similar variations, we cannot today predict what the exact distribution will be. Complex climate models are in general agreement on the magnitude of the global warming and on the much larger warming near the poles, particularly the margins of the Arctic Ocean. But the details of the heating in each region differ significantly from model to model. Similarly, these model experiments show increases in rainfall in some locations and decreases in others, but the agreement between different models is poor. This limitation of models will not be easily removed until our understanding of many aspects of the oceans, such as the vertical mixing of heat down from the surface and the relationship of ocean circulation to surface conditions, is improved. Even then, the simulation of transient effects will require long computer runs, perhaps a hundred years of simulated time, using models with improved ocean simulation, a finer mesh of points, and with gradually increasing amounts of infrared-trapping gases.

To summarize, climate models today are complex, sophisticated, and realistic. They are also coarse, approximate, and still evolving. So again we ask: Should we believe them when they tell us of a future that is both dramatically different from the present and possibly overwhelming? To answer this question we must gain confidence in the models by testing them in a number of ways.

The primary tests of the models involve comparisons with actual climates: comparisons of model calculations with current climate, comparisons with current climate change, and comparison with historical data sets. The simplest climate model can be compared only with the long-term average of the global temperature, while the ideal model should correctly predict the average temperature, pressure, precipitation, and other features not only for the earth as a whole but at each location on earth. It should also reproduce accurately the amount of year-to-year fluctuations around the averages. Existing models fall between these two extremes but are sufficiently detailed to allow a variety of tests.

First, one can ask whether the model seems reasonable. Are the internal details of the model correct. For example, are the energy flows of the right magnitude, do the fluctuations mimic what actually happens, are the winds correctly simulated? The best of modern models meet these challenges quite well and can therefore be subjected to the next test, in which we ask how accurately they simulate today's climate. Figure 20 shows a comparison of the climate, represented by surface temperature, as calculated in a modern model and as actually observed on earth during the last thirty years. The overall pattern is impressively similar; the climate in large regions is mostly correct. Even the errors that exist can be easily explained. For example, the temperature given in the model for England and Ireland in January is somewhat colder (270K) than is observed (280K). In this particular model the surface of the ocean is represented, but ocean circulations are not included, so the warm Gulf Stream is not avail-

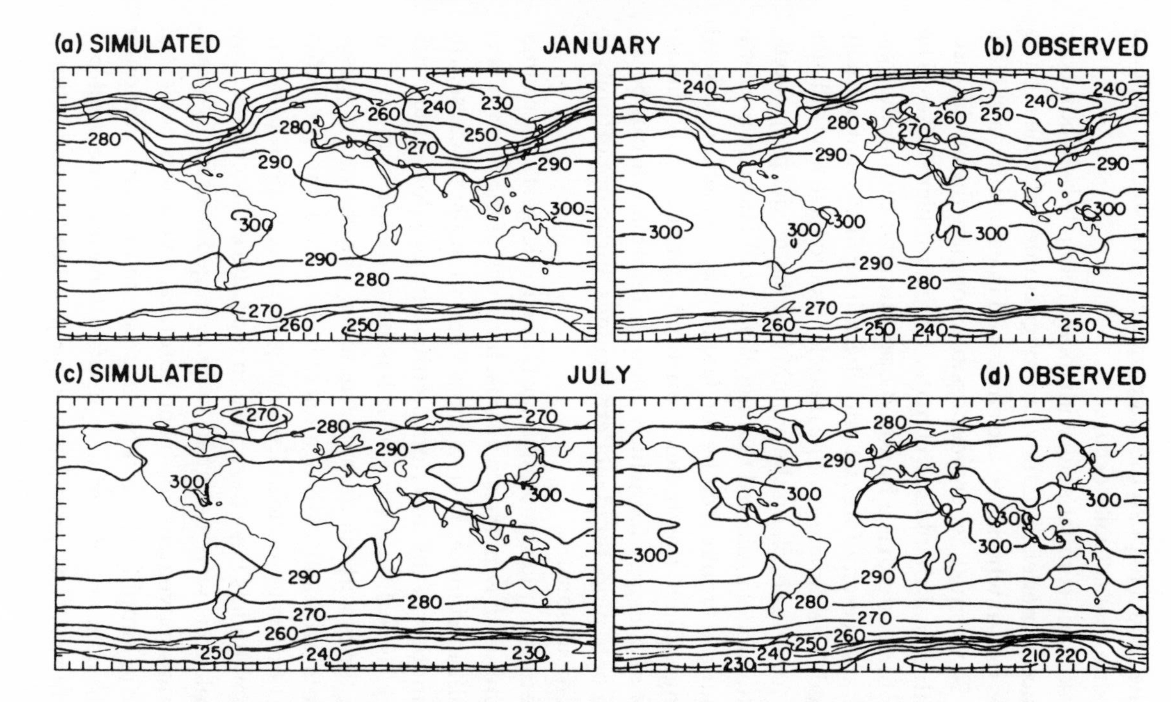

FIGURE 20. A comparison of the surface temperature as computed in a climate model and as observed. The top two diagrams are for the climate of January; the bottom two are for July. The contours show lines of equal average temperature and are labeled in degrees Kelvin. To convert approximately to degrees Celsius, subtract 273 degrees from these labels; the 280-degree contour passing through Cape Horn in the January simulation connects places with a temperature of 7° C, a few degrees above the freezing point.

able in the model to push the temperature of Western Europe up to the value that actually occurs.

These static comparisons with today's climate help build our confidence in models, but they cannot guarantee that the models will be as useful when they are employed in estimating a climate change. So we must, wherever possible, compare the model forecasts with actual changes in climate. Here a problem arises. The weather forecaster has a hundred opportunities during the year to make a three-day forecast and check it against what actually happened. The climate-change modeler would have to wait half a lifetime for one check on the accuracy of the model projections. And if the model projected that the earth would heat an intolerable amount, it would then be too late to take any action to avert the change. Thus, the modeler must turn either to slow climate changes of the past for which enough information can be assembled or to more rapid modern changes, for an opportunity to make comparisons.

The most rapid climate change is also the largest. The change from summer to winter each year is, for much of the earth, a very large shift in climate. It lasts for several months; its cause, the annual motion of the sun from south of the equator to the north and back, is well known; and it has been well documented in modern times with modern instruments. The change of season represents an ideal case against which to test the predictive powers of climate models. Curiously, it has only been in recent years that models have been able to reproduce this change. Earlier simulations were designed to be averaged over many years, so the sun's movement north and south each year was neglected in order to simplify the computer program and save computer time. But as more powerful computers have become available, and as interest has grown in seeing just how realistic climate models are, this seasonal simulation has been accomplished and the models have behaved well. In figure 20, the match between model and nature is as good for July as for January.[41]

Modern phenomena that are not as well understood as the seasonal cycle are also used to test models. El Niño, an occasional warming of the equatorial eastern Pacific Ocean surface, is associated with a characteristic pattern of weather over the North Pacific and North America. When an intense El Niño occurs, nearby changes are obvious—unusual winds and heavier than normal rain—but more subtle effects are observed over very large areas, including the North Pacific, the North American continent, and the Atlantic Ocean. Although what causes the eastern Pacific sea surface to warm in the first place is still not well understood, the best current models give a qualitatively correct picture of the large-area effects when the ocean temperatures in the model are changed in the appropriate way.

Verification tests using past climates depend on our ability to assemble an adequate picture of the past change and to know or guess the cause of the change. Some rather well described changes in climate have occurred since the retreat of the last ice age, and the transition from ice age to our present interglacial was an even larger change. Astronomical theory and observational evidence point to the slow changes in the earth's orbit around the sun as an important cause of ice ages and hence possibly of the smaller climate changes since. Scientists have been successful in piecing together a picture of the climate changes that have occurred since the retreat of the last glaciation, including not only some information about temperatures, but also miscellaneous data, such as various lake levels around the world. Lake levels are not simple climate indicators, since they involve rainfall (which contributes to the input of water to the lake), temperature (which governs, in part, the evaporation of water from the lake), and factors peculiar to each lake, such as the tendency of the water to overflow into another basin or river after it has risen to a certain level. But despite these difficulties, recent model calculations that use as their starting point the position of the sun in the sky nine thousand years ago give a moderately realistic picture of lake levels in Africa and the American Southwest at that time.[42]

In summary, the improvements made in model structure and sophistication in the last few decades and the various tests that have been performed using these models have produced a growing sense that, despite their remaining flaws, the models have reached a useful degree of realism and their projections must be taken seriously. The remaining weaknesses of the models account for the wide range of predicted values for future heating. But anywhere within that range the following statement holds true: If the concentrations of infrared-trapping gases in the atmosphere continue to increase as recent trends imply, the average temperature of the earth's surface will soon become greater than at any time in human experience.

But will the concentrations of these trace gases continue to increase as they have in the recent past? How much these gases will increase in the air depends on whether or not the world's societies decide to regulate the industrial and agricultural activities that release these gases into the atmosphere. We are not yet able to confidently estimate agricultural and industrial trends in the absence of policies governing trace-gas emissions, and, with one exception, no effective move toward global policy is yet in sight. The standard approach to estimating future concentrations, therefore, is to discuss today's rates of change of trace gases in the atmosphere and then speculate on factors that may modify them in the future. Each important infrared-absorbing gas presents somewhat different issues.[43]

Carbon dioxide was the first gas to be recognized for its importance to climate, and it has received by far the most attention. The concentration of carbon dioxide in the air a hundred years ago is estimated to have been between 260 and 280 ppmv; current measurements show a concentration of 350 ppmv, with a yearly increase of about 1.5 ppmv. This increase is almost certainly due primarily to the use of fossil fuels in electric power generation, transportation, home heating, and industrial processes. It is important to recognize that carbon dioxide is not a pollutant in the usual sense; it is a necessary end product of the full conversion of any fossil fuel into energy. Coal is mostly carbon; when it is completely burned, the end product, in addition to heat, is carbon dioxide. Liquid and gaseous fossil fuels are hydrocarbons; the gases resulting from complete combustion of these fuels are water vapor and carbon dioxide. We may learn to clean the sulfur from coal to lessen the problem of acid rain, but if we removed the carbon from coal, we would no longer have a useful fuel.

The amount of carbon dioxide released into the air is very large. In 1986, the use of fossil fuels put approximately 5 billion tons of carbon in the form of carbon dioxide into the air, and about half of that is still there. In other words, in one year the "average person" on earth accounts for the addition of one ton of carbon, or almost four tons of carbon dioxide, into the atmosphere. (There is, of course, no average person. A citizen of the United States contributes more than 15 tons a year of new carbon dioxide to the air, while in some countries the average is much less than 1 ton.) In any case, these large numbers discourage us from trying to design a way to catch the carbon dioxide as it streams from millions of smokestacks and tail pipes and storing it away somewhere.

Various techniques have been employed to estimate the amount of coal, natural gas, and oil remaining in the ground. These estimates are not perfect, but they are good enough to show that the remaining reserves of these fuels do not provide any useful limit to the amount of carbon dioxide that can be emitted. We could easily double or quadruple or eventually octuple the atmospheric concentration of carbon dioxide using known reserves of fossil fuels. Estimates of future emission must focus on what societies will do about energy use, not on how much coal, oil, and gas is available.

Different studies have produced quite divergent estimates of future fossil-fuel use. Some analysts, impressed with human society's long history of growth and industrialization and the length of time needed for new technologies and new fuels to become widely established, project continued expansion in fossil-fuel use. Others, noting the waste associated with much of the current use of fossil fuel, anticipate that economic considerations will force improved efficiency in the use of fossil fuel and a slow decline in its global consumption. The rate of increase of carbon dioxide emissions since the mid-1970s has been about 1.3 percent per year. Global use of fossil fuel had been expanding at more than 4 percent a year until the oil embargo of 1973 and the resulting large price increase slowed the rate to below 2 percent. Fuel prices have since dropped, however, and in the United States fuel consumption is on the rise again. At 1.3 percent growth a year, atmospheric carbon dioxide could reach 400 ppmv by 2015 and 600 ppmv by 2076. At 4 percent growth, the amount could be 400 ppmv by 2009 and 600 ppmv by 2039.

These estimates of future atmospheric concentrations depend not only on the rate of emission of carbon dioxide but also on the rate at which various processes on land and in the ocean can take up this gas and convert it into some form that is stored away from the atmosphere for a long time. In the ocean, most of the storage is accomplished through the circulation of surface water, saturated with dissolved carbon dioxide, to the deep ocean. This process is supplemented by the action of small creatures in the surface waters that use carbon dioxide to form their shells or bodies, then die and sink to the ocean bottom. On land, attention is being focused less on the storage of carbon dioxide by extra plant growth, but on the possibility that deforestation and other land-use changes may be adding to the increase in carbon dioxide concentrations. If a forest is converted into cropland, the carbon stored in the trees is eventually released as carbon dioxide into the air, either by the burning of the forest or the slower decay of the wood. Conversely, a new forest takes up carbon dioxide and stores it for as long as the forest continues at the same level of total biomass. Recent work indicates that deforestation releases only about a quarter of the new carbon dioxide appearing in the atmosphere, but that it may have been more important in the early buildup of atmospheric carbon dioxide in the nineteenth century.

Studies of carbon stored in trees also suggest that the nations of the world engage in massive reforestation, or undertake afforestation projects in which trees are planted on what has traditionally been grassland or cropland. Such forests could absorb carbon dioxide as the trees mature, and the resulting woodlands could then supply fuelwood indefinitely if they were harvested at the same rate as they grew each year. The amount of land required to take up a major fraction of the carbon dioxide emitted in the use of fossil fuels is, however, large. If we could plant 10 percent of all cropland and pastureland on earth to trees today, their annual growth would remove about one-fifth of the carbon dioxide we add to the air each year. Added to the one-half already removed by the oceans, these trees would slow, but not stop, the increasing concentration of carbon dioxide in the atmosphere, and they would do so only during the years in which they were growing to maturity.

Each of the other infrared-trapping gases presents a different problem. The simplest ones to study are the CFCs. The two main representatives of this group of gases, CFC-11 and CFC-12, are increasing in the air at a very rapid rate, nearly 5 percent per year (compared with about 0.5 percent per year for carbon dioxide). And unlike carbon dioxide, these gases do not interact with living material or the oceans, they have no nonhuman sources, and they are manufactured entirely for a rather narrow range of industrial and domestic uses. Because of the need to prevent further depletion of stratospheric ozone, these substances have already been the subject of an international agreement that may slow their growth over the next twenty years.

The most difficult gas to consider may be methane, whose concentration in the atmosphere is increasing at about 1 percent per year. This gas is released into the atmosphere mostly from anaerobic decay of biomass—in sanitary landfills, swamps, rice paddies, or the digestive systems of cattle and termites. Some methane escapes from natural gas wells and pipelines or from coal deposits; a small amount is the by-product of industrial processes. Methane is removed from the air when it reacts with the hydroxyl radical (an unstable molecule made of one atom of hydrogen and one of oxygen). The cause of the rising amount of methane in the atmosphere is unknown, but scientists speculate that the increase in paddy rice and cattle production, growing numbers of termites in the tropics, heavier use of landfills for urban waste disposal, and the competition for the hydroxyl radical by carbon monoxide released from automobiles and industry may be responsible.

Nitrous oxide, another infrared-trapping gas that is veiled in uncertainty, is also a complex by-product of biological activity. The use of nitrogen fertilizers in agriculture may be causing the increase in the atmospheric concentration of nitrous oxide observed in recent years. It is currently increasing in the atmosphere at a rate of about 0.3 percent per year.

These are the major infrared-trapping gases emitted into the air by some mechanism and now known to be increasing in concentration. One more gas must be mentioned, however: tropospheric ozone. Ozone is not directly emitted into the air but is created in the lower atmosphere by a complex set of chemical reactions. Ozone, of course, is already famous, first as the gas in the high atmosphere that absorbs ultraviolet light from the sun, thereby reducing the amount of sunburn humans experience, and second, as the gas that, near the earth's surface, helps acid rain damage the forest. But ozone also absorbs infrared radiation effectively, and thus it is also classified as a climate-heating gas. It is thought that the average concentration of this gas is increasing in the lower atmosphere. If this increase is anthropogenic, it is probably attributable to releases into the air of nitrogen oxides and hydrocarbons—by-products of a wide range of industrial and agricultural activities, from driving cars to raising livestock. These gases add to the reaction chains that create ozone and so increase its concentration in the lower atmosphere. The connection between carbon dioxide, CFCs, and ozone and our industrial and domestic activities, and the relationship of methane and nitrous oxide to the world's agricultural activities, mean that the release of these gases is thoroughly intertwined with our present techniques of providing for our health and prosperity. Minor adjustments in technology, foreign aid, or Third World–debt-management plans, among other areas, are not likely to slow the climate change by very much. More fundamental shifts must be undertaken.

Clearly, it is more difficult to estimate the possible future concentrations of methane, nitrous oxide, and ozone than those of CFCs and carbon dioxide. But some sort of projection is needed. In particular, we need to know whether, with emissions continuing at possible and reasonable rates, large climate changes could occur. In order to make such a projection we can take the present rates of increase for these gases (but using lower values for the CFCs, because the Montreal Protocol, even with its compromises and delays, will surely slow the rates somewhat) and compare their calculated effects with those of carbon dioxide, assuming that fossil-fuel use continues to rise at its present rate. The total warming thus computed is about twice that for carbon dioxide alone.

Such a rapid change in climate will have many consequences, some of which have been mentioned earlier. Sea level, which rose 100 meters or more during the retreat of the ice age, can be expected to rise even further. The pattern of rainfall and evaporation will change, forcing dislocations on various ecosystems. Natural forests will be stressed and agricultural patterns will have to be modified. One might think that the prospect of such a rush into a threatening future would have set world leaders on a more cautious course and brought about a more thorough exploration of the new situation. That this has not happened is of course due to the difficulty of slowing anything so basic to the current world economy as fossil-fuel and food production.

This difficulty has led to a reexamination of the mathematical and physical underpinnings of the models and to the testing of the models described earlier. It has also led to a careful examination of each of the assumptions employed in the models and in the process of foreseeing severe impacts, to determine if some phase of the calculations might be producing unnecessarily alarming results. We can all be forgiven for hoping that perhaps some aspect of the climate system will automatically limit the degree of change, or that some overlooked process in the ocean or biosphere will stabilize the amounts of carbon dioxide, methane, nitrous oxide, and ozone in the atmosphere. Perhaps we will even decide that the rapid heating will do more good than harm. But so far the investigation has only made the situation appear more critical.

A decade ago there was controversy about the size of the temperature increase. Some of the simple models showed much less warming than that calculated by Arrhenius or the values being discussed today. But careful examination showed that those models neglected some critical aspect of the problem, and they have now been superseded by a vastly more complex and meticulous accounting of known effects in the atmosphere.

But is there not an ice age on the way? The slow changes in the earth's orbit around the sun are now known to be associated with the ebb and flow of major glaciations. Over the course of a few million years the earth has seen periods when the temperature was as warm as it is today that lasted perhaps twenty thousand years, followed by longer periods of ice and cold. The earth emerged from the last glaciation about fourteen thousand years ago, and scientists know no reason why the earth should not enter another cold period in the next several thousand years. However, the transition to a glacial period is slow—a few hundredths of a degree per decade—and in the next few centuries it would be completely canceled out by the more rapid heating caused by infrared-trapping gases.

Scientists and others have proposed various secondary processes in the climate system that might offset most of the warming. A number of secondary processes are already included in all the numerical models. For example, as the earth's surface and lower atmosphere become hotter, more radiation is sent upward, more heat escapes through the infrared-trapping blanket, and the amount of surface heating is thereby limited. An example of a process that makes the heating greater occurs in the reaction of the high-latitude snow-ice line to warming. The warming of the atmosphere initially causes the ice and snow line to melt back toward the poles, uncovering earth, which is darker than snow, and thus increasing the amount of heat absorbed from the sun and producing further warming.

Clouds present a difficult challenge to the creators of climate models. Clouds reflect sunlight away from the earth, and they trap heat near the surface of the earth. Thus changes in clouds that might accompany a climate warming could strongly influence the amount of human-induced climate change. Press accounts frequently speculate that the increase in atmospheric moisture that should accompany a climate warming might result in more clouds, more reflection of sunlight, and hence a slower climate heating. However, the current group of models does not show a decrease in the warming trend that can be attributed to changes in the clouds. These same models simulate the change of season from winter to summer fairly well, even though both moisture and clouds change, thereby raising one's confidence in the models' treatment of clouds in the models. The models are indeed sensitive to how clouds are incorporated in the calculations: differences in the treatment of clouds account for most of the difference in heating estimated by the various model calculations. But all the models project a rapid climate heating when combined with extrapolations of today's emissions.

Other speculation has focused on factors not now included in the models, especially processes that might change the rate of emission of infrared-trapping gases or the rate at which such gases are removed from the air. The greatest uncertainty surrounds what society will decide to do about its rate of emission of infrared-trapping gases. Because the models are being used to advise governments on the impact of emitting these gases at various rates, it would be circular to try to include society in the models, even if we knew how. There are, however, processes not involving human activities that could modify emission or removal rates, and these have been examined in the hope of finding some automatic relief from the projected heating. For example, searches have been mounted for possible ways whereby marine organisms or forests might accelerate their uptake of carbon dioxide and thus remove more of it from the atmosphere. The simplest effect in the ocean, however, works in the wrong direction.

As the surface water warms, it can dissolve less carbon dioxide, and this observation indicates that ocean uptake would decrease in a warmer climate. But predicting whether the transfer of CO_2 to the ocean will actually change as the climate heats depends on a better understanding of how ocean circulations will be modified. Other ocean processes may influence these events but again in the wrong direction. Deposits of methane complexes at the bottom of shallow arctic seas may emit methane as the seas get warmer and so heat the earth even faster. Some plants will grow more rapidly because of the additional carbon dioxide in the air, but most plants die back or drop their leaves each autumn, thereby returning much of the carbon dioxide to the air and minimizing the amount of extra long-term storage of carbon dioxide away from the air.

There remains the possibility that the total amount of wood and organic material in mature forests and their soils may increase, because trees will have more carbon dioxide available with which to carry out photosynthesis. Balancing such an effect, even overpowering it, would be the increased rate of decay of forest carbon by the higher average temperature and an accompanying increase in bacterial activity. Added to the carbon dioxide emission from this source would be emission of both carbon dioxide and methane from the decay of material that is currently frozen in high-latitude bogs and peat deposits.

But the most compelling reason for discounting the speculation that nature will somehow cleanse the atmosphere for us is that we continue to measure steady increases in the concentration of carbon dioxide in the air, year after year. Again we are reminded that human releases of carbon dioxide and other gases into the air have grown well beyond what the earth can promptly remove.

In recent years a new idea has been put forward that has persuaded some people that the climate heating may be automatically controlled. The Gaia hypothesis, as this idea is known, states that in the long course of evolution plants and animals have developed not only the capability of surviving in their surroundings, but also the ability to control their surroundings to improve their survival chances. So, the argument goes, if the earth starts to warm, living things will so modify the color of the surface, the composition of the atmosphere, or other global characteristics in order to restore the temperature to its optimum setting. The Gaia hypothesis has generated considerable scientific interest and debate, but it will not help us with the climate-heating problem. During the Cretaceous period, when dinosaurs roamed, the earth was much hotter than now; during the peak of the last ice age it was much colder. If Gaia controls temperature, it does so very loosely, allowing changes that could devastate modern civilizations if they occur too rapidly.[44]

Finally, some people have tried to find hope in the positive consequences of the predicted change in climate. Some nonscientific writers are given to fantasizing that longer growing seasons and warmer temperatures will speed the growth of trees in the Northwest Territories, Siberia, and other cold regions, forgetting that plants are adapted to the conditions where they grow and that a longer warm season can thus be as damaging as a shorter one. Many people express the hope that the shifts in rainfall that will accompany the climate change will bring more water to arid regions, especially in Third World countries, and thereby alleviate problems there. Such sentiments should be viewed with caution. The social and political systems in these countries, and the interaction of their economies with the world economy, have created a situation that is poorly adapted to the current climate. City dwellers demand lower food prices than farmers can tolerate, for example, and more and more people try to live in areas that cannot reasonably be expected to support such numbers. Pinning one's hope on changes in rainfall patterns may well be a way of avoiding the deeper problems of overpopulation, maldistribution of wealth, misuse of commercial power, and political instability.

Agriculture in the industrial world is of course more flexible than in less developed nations, and it is possible that prosperous farmers may be able to take advantage of changes. Some plants will need less water than they do now because of higher carbon dioxide levels in the air. This change occurs because the plant, supplied with more carbon dioxide than it needs, reduces the size or number of small openings in its leaves that are used to take up carbon dioxide. As a result, water loss through those same openings diminishes. Partly offsetting this advantage will be the fact that some weeds are able to use the extra carbon dioxide more effectively than some crops. And in unmanaged forests and grasslands the different reactions of the various species to added carbon dioxide will alter the balance of species in ways that cannot be predicted today.

The reaction of plants to changing conditions illustrates one increasingly important feature of the problems we humans have wandered into. Modern agriculture is riding the crest of decades of technological success. Agricultural research has provided solutions to many of the problems we face in providing a growing population with food and fiber. Selection and crossbreeding of crop strains, fertilizer applications, and the use of chemicals to inhibit weeds, kill insects, and restrain parasites and diseases have produced both a green revolution and the confidence among agricultural experts that no obstacle is too great to overcome. Until recently, therefore, predictions of undesirable species shifts would be met with suggestions for applications of herbicide or nutrients to restore the balance. But in the 1990s much of the focus will shift to figuring out how to decrease the use of agricultural chemicals. Chemicals have produced favorable results in the cultivated fields, but, like DDT in the 1960s and

1970s, they have also been accumulating in woods and waterways and causing trouble. Regulatory agencies are beginning to propose limits for "nonpoint source pollution" (that is, pollution coming from broad regions rather than from a single drainpipe at a chemical factory or oil refinery). Even in their primary uses, chemicals are far from perfect. At first pesticides kill insects or weeds, but in the end they generate new resistant strains, thus requiring a new round of chemical research, product development, and testing.

The striving for more accurate, more complete climate models has accelerated in recent years as increasing numbers of skeptical policymakers have been forced to consider whether action is needed. Increasing numbers of scientists have been drawn into the search for better simulation techniques, improved global and regional data sets for model tests, and more sophisticated methods of estimating the impacts of climate change on human activities. This search will continue, but we can evaluate the progress to date:

- The soundness of Arrhenius's basic idea has withstood all assaults.
- The realism of the simulations produced by models has increased the scientific community's confidence in them.
- The preponderance of the effects of a major heating will be harmful to people and to natural ecosystems.

In summary, the best efforts of the world's scientists foresee a rapid heating of the climate, and a vigorous search for reasons not to be concerned about this change has so far failed.

6. Just One Problem

Acid rain, ozone depletion, and climate heating—the three most highly publicized aspects of the rapidly changing composition of the atmosphere—are more than just related. A single, larger problem underlies them. They are all three the result of the impact of human activities on the earth, which now equals and even exceeds the influence of large-scale natural forces. Whereas humans once were overwhelmed by natural phenomena such as volcanoes and hurricanes, droughts and blizzards, today they dominate their physical and biological environment. All across the Northern Hemisphere, the proportion of human-produced sulfur dioxide in the air is several times greater than that produced by plant life and volcanoes, and its effects on soils, plants, and fish have outstripped the cleansing ability of the slowly evolved processes that maintained the biosphere during preindustrial times. In the stratosphere, increases in the concentration of synthetic substances have begun to destroy the ozone layer. And the lower atmosphere now holds infrared-trapping gases in sufficient quantities to disturb the long-standing balance of gains and losses of heat by the earth's surface.

This statement surprises some scientists and alarms some

political leaders, but it leaves most people unimpressed. Of course we dominate our surroundings, most people will say; what is surprising or disturbing about that? We keep our temperature moderate with houses and machinery, make travel possible with roads and airports, lessen floods with dams and levees, eliminate harmful or irritating insects with chemicals, and convert land and divert water to produce the food we require. This dominance is designed and desirable, they will say. And then they may add, "Would you have it otherwise?" In this chapter I will attempt to answer this fundamental question.

People have been modifying their surroundings and creating wastes for as long as there have been people. The same is true of other living species: trees drop leaves, foxes dig burrows, beavers cut trees, algae turn lakes into bogs. Our usual picture of the preindustrial and pre-agricultural earth is one in which many species competed for space and resources, the waste products of one species were utilized by another, and the population of each species was held in check by disease, accidents, predators, but primarily by a limited food supply. It is likely that this arrangement was far from static. A warm century or a cold decade could shift the balance of species; a windstorm or a forest fire caused by lightning could modify the nature of a forest; a new land bridge between islands or continents created by a low sea level could bring new species into an area and upset ecosystems for all time to come. But throughout these kinds of changes, no one species dominated everywhere or for long because the competition was just too lively.

Then, rather recently, humans were able to shake off some of the constraints that restrict other species and run to the head of the line. We learned to cultivate crops and to irrigate, fertilize, and protect them from pests so that the amount of food produced on each plot could be increased. We organized ourselves and specialized our skills in order to produce what we needed more effectively, and within the last few centuries we learned to use fossil fuels to drive and advance our technology. Our skills in producing what we want, as well as our numbers, are still increasing at a breathtaking rate, and with them the variety and amount of modification of our surroundings, as well as the number and height of our piles of waste.

There is a paradoxical aspect to this situation, however. Yes, we have freed ourselves of constraints that keep other species in check. But we are still largely dependent on other species and ancient ecological systems for our health and progress. The plants and animals we eat evolved with us or before us and, though selected and grafted, inbred and crossbred, still carry the genetic information acquired in the slow trial and error of evolution. Fossil fuel, which drives the civilizations of the world, is created (in ways we do not completely understand) from the remains of plants and animals processed under the earth's surface. Our houses are mostly made of wood, or stone and sand stuck together with heated limestone—limestone that was laid down in ancient seas as the shells of small creatures accumulated, bit by bit. And around us all is the atmosphere, upon which we depend each minute of our lives, which was created in the first place by the early bloom of life on earth and is renewed daily by trillions of members of a million species, most of whose lives we have yet to study.

Glimpses of this paradox disturb us, now and then. The end of our fossil-fuel reserves is in sight; what then? The genetic library of plant information is being burned down by the invaders; will we enter a dark age of food and medicinals? The atmosphere is out of kilter; the waste products of one species overpower the cleansing ability of all other species. Where will it all lead?

By and large, people do not spend much time asking these questions; we are not greatly worried by changes in the world we live in. The reason seems to be that we have quietly assumed that the future will take care of everything. If the Dutch managed to reclaim a bit of the sea floor for farmland; if modern medicine could eliminate smallpox, a serious disease, from the earth; if antibiotics can control other diseases; and if floodwaters can be retained in reservoirs, then certainly we are skillful enough to manage everything.

Our tendency to free ourselves from the world as we found it was expressed more than fifty years ago in an editorial about the weather in a Boston newspaper:

> Weather forecasting will become increasingly unimportant in the future. Of course a lot of people will still like to know whether to take an umbrella and rubbers when leaving the house in the morning, but, more and more, it won't make much difference if they take or forget them.
>
> Fifty years ago much land in the tropics was devoted to growing indigo. The crops were often severely injured by droughts. Obviously it would have been a great help to the indigo growers if droughts could have been predicted well in advance. But today such long-range forecasting would be of little value, because virtually all indigo is now made from coal tar in factories.
>
> So a hundred years from now it is quite possible that the forecast of a heat wave will be only of academic interest to most persons. We shall probably be working in air-conditioned offices and factories, sleeping in homes where we may select our favorite temperature, and traveling in trains, automobiles and airplanes, all of which will be protected from the outside weather. Then only farmers, incurable out-of-door friends and a few sentimental diehards will even bother to read the weather reports![45]

Sometimes this point of view is accompanied by the implication that it is our duty, not just our destiny, to manage everything.

Thirty years ago, a U.S. senator scolded the executive director of the Sierra Club, who had come to Washington, D.C., to testify against the authorization for the construction of a certain western dam: "I would like to remind you that not only did God make the earth, He made it for man, and one of the first commandments that He gave him . . . was to multiply and replenish the earth and subdue it."[46]

It only takes a brief search through today's magazines and newspapers to learn that this view still prevails: updated, softened, secularized, but unchanged. In a book published as a tie-in to a television science special, we read: "Yet if we learn to know the climate system well enough, its exquisite sensitivity may become a blessing to us. We may then be able to make selective improvements in the climate or arrest its decline . . . there may come a time when humankind drives climate, instead of being driven by it."[47] Other comments of this sort were heard at recent meetings concerning climate heating and other features of rapid global change. An economist participating in a discussion on the impact of climate change dismissed the problem impatiently, remarking that surely scientists could figure out how to make the air somewhat less transparent to incoming sunlight and thereby balance the increased infrared trapping. This suggestion was echoed at another meeting by a participant who asked whether scientists could not arrange to explode a nuclear bomb once a month to create just enough "nuclear winter" to balance the "greenhouse summer."

Like these writers and speakers, most of us are technological optimists. If fossil fuel disappears, surely nuclear fusion or giant solar mirrors in space or something we have not yet thought of will arise so that we can continue to meet our needs. We can fire nuclear wastes into space or bury them in deep ocean trenches. As ecosystems disappear, we can replace them with crops we want or beautiful formal gardens or greenbelts; we will probably need the room anyway for our growing population. If the climate heats and the sea level rises, we can shift crops to new locations or, better yet, design new crops in order to produce food wherever it is needed, perhaps in greenhouses or larger, more elaborate controlled environments that we will need anyway in space and on the moon and Mars. In addition, we can build seawalls to give our cities the time to migrate inland and perhaps design underwater cities where the climate never changes and many problems are easier to solve.

And perhaps we would be right. Perhaps we stand at the beginning of the next chapter in the saga of human domination of the universe—the chapter in which we free ourselves of dependence on the ancient, evolved systems of the earth's biosphere, atmosphere, ocean, and surface. It is a tantalizing thought. The list of recent technological successes is impressive; not only do we solve problems and invent things, we also study how we solve problems and thus speed up the next round of progress. Our readiness to free ourselves from the earth's remaining constraints is nowhere more clearly symbolized than in two developments of the late twentieth century: spaceships that allow us to leave the planet and laboratories where we can build new living species from the blueprints of old ones. With these techniques in hand, it is hard to be bold enough in imagining possible futures. Can we develop trees and other plants that are resistant to air pollution to replace those now fading away? Or why bother, since we can probably put new kinds of plants on Mars that will begin to create a useful, heat-trapping atmosphere on that barren planet and provide ourselves with an alternative world? Can we alter future humans so that they can live and work in a wider range of environments?

Even at the level of more familiar technologies, the future looks bright. It is not too great a step to visualize (much as the Boston editorial writer did fifty years ago) previously "useless" deserts covered with climate-controlled housing-workplace-recreation complexes, powered by some form of solar energy, where news, entertainment, access to data bases, and electronic mail-order shopping are offered through the next generation of television devices, and which one may reach by suborbital air- or space-craft traveling at very high speeds.

This now familiar list of possibilities is usually put forward with only one qualification: these things are feasible only if we can manage not to blow ourselves up before we get started. But there is accumulating evidence that another qualification should be added to the first. To continue successfully along the current path requires not only that we replace the evolved systems on which we depend, but also that we be wise enough to do so in the right sequence and completely, so that at no time during the process are we left without food to eat, air to breathe, and strong governments to keep the peace. This qualification is needed because the increasing domination of natural systems by our expanding technological society also means that in addition to freeing ourselves from our dependence on these systems, we are also causing their gradual disappearance. We will need to replace the accumulated "wisdom" of the interconnections between air, earth, water, and species with our own intelligence, diligence, and management skills.

There are many uncertainties along this path. The first of these is that technological progress is not always as smooth as we would like. The hopes of thirty years ago, for example, that nuclear energy "too cheap to meter" would meet all our energy needs have been replaced with fears about supply, prices, pollution, and international conflict. Photovoltaic cells have advanced more slowly than hoped. Although they work, they are still too expensive to contribute substantially to the world's existing network for the generation and distribution of electricity. Like nuclear fusion, batteries suitable for electric cars are always just around the corner. The uncertainties are not limited to the energy sphere. With repeated use, pesticides create new pests, and fertilizers enhance the need for fertilizers. These are straightforward technological delays or failures. They produce no great harm as long as we have time to try something else, but they caution us not to cling to the simple belief that we can always come up with the correct new technology just when we need it.

A larger group of failures involve technological innovations that fail to solve the problems they are designed to solve. Many years ago a classic study of the effectiveness of flood-control projects in a major river basin showed that the structures erected increased, rather than decreased, the total damage due to floods. People, encouraged by the flood-control work, built more and more valuable buildings in the floodplain, so that when the now rarer flood did occur, the damage was greater.[48] Some more recent studies of the green revolution show a related chain of events. In order for high-technology crops to be grown, small farms had to be aggregated into large holdings and hand labor replaced with machinery. The net result was more crop production but also more hunger, as small farmers found themselves reduced to day laborers or migrated to large cities with insufficient income to buy the magical crops so produced. This agricultural example has a global counterpart. Year after year, our technological skills have succeeded in producing, on the average, more and more tons of food and fiber from each square kilometer of farmland until the amount of protein and calories now produced worldwide roughly meets the minimum dietary needs of everyone on earth. Yet, year after year, the number of starving people grows larger and larger.[49] These failures suggest that our ability to create the social institutions needed to make proper use of our technology lags behind our imperfect ability to invent new technology in the first place.

A third type of technological failure is illustrated by the three atmospheric problems described in this book. Our ability to foresee unintended, harmful side effects of a newly introduced technology is severely limited. This limitation arises in several ways. The first people to release small quantities of a CFC into the air may never have even heard of the stratosphere or the chemistry of the ozone layer; their tests showed the substance to be "for all intents and purposes" inert. The small quantities released would be so diluted by the vast global atmosphere that even unanticipated damage would be reduced to harmless and undetectable levels. But once the substance proves useful, as coal or CFCs certainly did, the amounts we release rise a thousand- or millionfold, and the benefits we derive from the substance allow us to tolerate and ignore the first signs of harm. We thus set ourselves up for serious problems when the amounts released increase even further.

The occurrence of such failures is adequate cause for asking whether we are ready to embark on global management of the soil, the air, the oceans, and those species we decide should survive. For a people who are not yet able to describe fully the workings of a tree or the chemistry of an ocean or to put available food in the hands of those who need it, it would be perilous to undertake the design and maintenance of a complete life-support system for everyone.

The complete reliance on our technological and management skills to solve future problems presents another difficulty. As solutions involve increasingly complex technologies and higher and higher costs, more of the world's people are left behind.

As the sea rises, for example, various countries will be affected. Land that is high above sea level, with cliffs for shorelines, will be affected only to the extent that installations such as harbors need to be adjusted. The Netherlands, on the other hand, will find it necessary to raise the height of its system of dikes by as much as the sea level rises if it wishes to maintain the same allowance for high storm tides that it now has. But what will Bangladesh do? Any atlas will show how close to sea level much of that country is; any geography text will describe the great numbers of people who inhabit and farm land not far above today's sea level; and newspapers report the heavy loss of life when hurricanes inundate the lowlands. But Bangladesh, which has far fewer resources than the Netherlands, is unable to afford the extensive network of structures required to limit the damage that storm-driven waves on a higher-than-average sea level would cause.

Farmers in Third World countries generally have fewer options for change and adaptation than those in developed countries, yet changes in temperature, increases or decreases in rainfall, and adjustments in the world market prices will all affect them as well as farmers in more prosperous lands. In general, Third World countries will have much greater difficulty adapting to rapid change than the developed world, even though the climate change is being driven mostly by gases that the industrial countries are emitting into the atmosphere. The rapid changes in the atmosphere may therefore intensify disparities between rich and poor countries and create new tensions among the world's peoples. At a recent meeting in India to discuss the impact of climate change on developing countries, a representative from Bangladesh startled the audience when he said that he assumed that the rich countries, now flooding the air with heat-trapping gases, would be willing to accept the masses of refugees that would be generated as his country was gradually inundated.

The course of development we are on is a risky one, demanding knowledge and skills we may not have and leading to consequences that we may find difficult to tolerate. At the most basic level, we need to ask whether the assumption that we can manage the earth and all life upon it, is leading us to the kind of world we would like to live in. An earth diked against a rising sea, with an atmosphere darkened to reduce sunlight at the surface, is not one that would appeal to most people. Of course, young people, born into such a world, would not be disappointed at a darker sun, nor would they miss the vanished species. It may well be that the real tragedy of air pollution is not that it will kill us, but that it won't: we will adjust to the changes and forget the possibilities that disappeared with the earlier world.[50]

7. The Other Path

There is an alternative path of development. It is a difficult one to contemplate, more difficult in many ways than the path of continuous rapid expansion, but it deserves consideration. This path requires that we lessen our impact on the atmosphere enough to slow the changes and allow time for people to adapt in the normal course of their activities. With more gradual changes and a reduced impact, we could continue to rely on the slowly evolved systems of air, water, and living things that have provided the environment within which human civilizations developed and that continue to support life in marvelously intricate ways.

The problem, of course, is how to achieve such a change in direction, what steps do we take to reduce the human impact on the earth? Some parts of the answer are known; others would require great effort and imagination—even a more gradual approach requires its own degree of technological optimism. Furthermore, the task is so vast that there is no hope of a single, simple answer. No new source of energy will suddenly turn things around: the use of fossil fuel did not slow human impacts, neither did nuclear energy, and neither will the next new source. The reduction of human impacts involves all aspects of our civilization: the number of people on earth, what each one of them does, and how he or she does it. A strategy with a fighting chance for success must call for stabilizing, perhaps even reducing, the world's population and reorganizing economic activities so that each person produces less waste and less impact on the earth and its systems. To be successful, such a strategy must also encourage lives of opportunity, variety, and challenge.

The atmosphere and its interaction with people and other parts of the biosphere are the focus of this book, so the recipe given here for finding a path with a smaller human impact on the earth will center around stabilizing the composition of the atmosphere. But almost every step envisioned for reducing human-induced changes in the atmosphere involves facets of industry and society that are not directly concerned with the air. Some carbon dioxide, for example, is released into the atmosphere when forests are cleared to make cropland or pasture, mostly in tropical countries. Any reduction in these emissions would require that we address issues of poverty, land reform, population pressure, and Third World debt. Indeed, had the focus in this book been on some global problem other than the changes under way in the atmosphere—the permanent loss of species in the tropics; political instability in the Middle East, Africa, and elsewhere; or the danger of oil spills on beaches—the recommendations would bear a strong resemblance to those for stabilizing the composition of the atmosphere. Solving any major global problem goes a long way toward solving them all.

Our problems stem from global increases in the concentrations of carbon dioxide, chlorofluorocarbons, methane, nitrous oxide, and a long list of other substances, mostly industrial chemicals, in the atmosphere, and regional increases in the concentrations of sulfur dioxide, nitrogen oxides, ozone (in the lower atmosphere), hydrocarbons, and some other substances. Many of these substances—and most of the very important ones—are related to fossil-fuel energy production and use. Therefore, the first step in finding a more gradual path must be to reduce our consumption of fossil fuel.

A direct approach to using less fossil fuel is unappealing. Traveling less, keeping homes and businesses colder in winter and hotter in summer, reducing street lighting—all would save fossil fuel but would also create problems or at least produce considerable public resistance. In the Third World, where lives are already sorely constricted by the lack of energy resources and hence opportunity, such steps would be in direct opposition to development goals. Thus, two other options for using less fossil fuel need to be examined: replacing fossil fuel with other energy sources and using fossil fuel more efficiently. Replacing fossil fuel with other sources is clearly a long-term requirement. Even without atmospheric impacts, we would someday exhaust all readily obtainable petroleum, gas, and coal. But for immediate purposes, improvements in energy efficiency are vital, and it is fortunate that the part of an overall strategy that is best understood addresses precisely this area.

Energy efficiency is sometimes confused with energy conservation, which got a bad name in the mid-1970s, when it was likened to "freezing in the dark" by representatives of oil interests. The true potential for efficiency improvements is far greater than the savings realized by simply turning down the thermostat or not leaving the lights on needlessly. If the light we need is provided by a 100-watt bulb, whereas an equally bright 20-watt bulb is available, switching to the lower-wattage bulb could save 80 watts whenever the light was on, with no loss of illumination to the consumer. Similarly, heating a house with a natural-gas furnace that extracts 90 percent of the possible heat from the fuel is just as satisfying, but less demanding of fossil fuel, as heating the same house with a furnace yielding (as is common today) a much smaller fraction of the available energy. If such efficient technologies are widely available, and if the cost of substituting them is reasonable, then savings in energy, particularly fossil fuels, will be practical components of a low-impact path for human civilizations.

Numerous modern studies show that these technologies do exist and that, for the United States, the goal of maintaining current levels of activity on as little as half the current amount of fossil fuel is technologically possible.[51] Furthermore, for many of these energy-efficient technologies, the savings in fuel costs would more than pay for the new equipment required. Thus, we have the opportunity to slow the progression of acid rain, forest damage, and climate heating while spending less on heating and cooling, manufacturing and processing, cooking and traveling, and indoor and outdoor lighting. These steps, incidentally, would also lessen urban smog, minimize coal-mining impacts, hold down the number of oil spills, improve industrial competitiveness, reduce national spending for energy imports, and eliminate a number of arguments over where to put the next power plant. Stated in this way, the possibilities sound almost too good to be true, but in fact Japan and Western European nations already manufacture goods using only half the energy the United States requires to make the same items, while maintaining a high standard of living. The United States has made a modest start in the direction of higher efficiency. For a decade after the 1973 oil embargo, per capita fuel consumption by Americans dropped steadily. The gross national product also increased by a third during the same period, and population increased by almost twenty million people. Improvements in energy efficiency allowed this growth to take place with no increase in total energy use.

But stabilizing fossil-fuel use alone, as the United States did during those years, is not enough to stabilize atmospheric carbon dioxide. In each recent year, human activities raised the amount of carbon dioxide in the air by about 3 billion tons of carbon, and with constant fossil-fuel use, we can expect similar increases to continue to occur. In order to stabilize the concentration of CO_2 in the atmosphere, we will have to go well beyond holding energy use constant and make changes that will substantially reduce the amount of fossil fuel we consume each year. Just how much we need to decrease emissions is not known, but a rough estimate, based on the idea that the oceans, in response to the current high concentration of carbon dioxide in the air, will continue to take up and store several billion tons more than they did in preindustrial times, suggests that we must aim at cutting emissions by half or even more. Further studies may refine this estimate in the coming years, but it matters little if the CO_2 emissions need to be reduced to 30 percent or 55 percent of today's value: the initial actions required are the same.

Coal is the fuel of greatest concern. Its mining releases methane, its transport requires energy, and its use produces carbon dioxide, sulfur dioxide, and nitrogen oxides. As a fossil fuel, its use produces twice as much carbon dioxide as natural gas for the same energy production. Because most of our coal is used to generate electricity, efficient electrical equipment is a top priority in reducing harmful emissions to the atmosphere.

The measures we must take touch a wide range of human activities: raising standards for home and office insulation, replacing electric resistance heating with heat pumps or natural-gas furnaces, replacing old air conditioners and other appliances in homes and old electric motors in factories with high-efficiency new ones, and replacing standard incandescent light bulbs with high-efficiency sources. Each of these steps can reduce our need for electricity, and hence for coal, by substantial amounts. Take the case of home refrigerators. In the United States, refrigeration accounts for about a fourth of all residential electricity use. Replacing the average home refrigerator with the most efficient one now on the market would save one-half to two-thirds of the energy use, accomplish the same amount of food storage, pay back the extra investment through energy savings within a few years, and, if adopted nationwide, eliminate the need for a dozen or more large coal-burning power plants. The technology for all these changes already exists, and the techniques for accelerating the rate of replacement are well known: advertising the savings that the higher-efficiency equipment makes available, issuing government standards for efficiency, taxing inefficient items or coal, and requiring the government to set an example in its operations and contracts.

Such changes are not without their problems. One powerful objection voiced against improved electrical energy efficiency as part of the cure for acid rain and climate heating is that it will throw coal miners out of work, as indeed it will. It is in the nature of our society for its technology to change rapidly and for one commodity to replace another at frequent intervals. The response to this aspect of our high-tech life cannot be to guarantee each person the same job for life, but must instead be to provide the social systems needed to smooth the transition from one job to another for individuals and from one business to another for communities. Further, such transitions will be easier to effect in an active, productive economy, and, in the long run, eliminating waste and reducing environmental damage will surely contribute to making our economy stronger. But our political system does not always take the long view, and so the issue of coal miners' employment continues to dominate the debate over acid rain. It is often representatives of the coal industry who argue that climate heating is "just a theory."

The employment impact of proposed steps to reduce energy use has been estimated by several studies, which tend to show that efficiency measures create as many jobs as they eliminate.[52] But the authors of the studies caution that their estimates are uncertain because of the many indirect effects any major national policy change has on the job market.

Petroleum is more convenient than coal for many uses. It is easier to process into a variety of forms; it can take the form of a liquid or a gas and therefore is easily transported through pipelines or carried as fuel in vehicles. Except for Eastern Europe, the industrial world uses more petroleum-derived energy than coal-based sources, and examination of petroleum use shows, just as in the case of coal, major opportunities for savings. With about a third of all energy in the United States used to power vehicles of various kinds, the energy efficiency of the most common vehicle, the automobile, is of critical importance.

The average car on the road gets about 19 miles per gallon in the United States and 24 miles per gallon in the rest of the world. Cars with fuel economies of up to 50 miles per gallon are on the market, while prototypes have been tested that get nearly 100 miles per gallon. However, the availability of these improved vehicles does not guarantee that the world will soon use less fossil fuel. Automobile manufacturers know that consumers do not look primarily at fuel economy when buying a new automobile; such things as acceleration, roominess, safety, and style will generally take precedence. And even though a high-efficiency car saves money for the buyer, it does so only if the lower fuel costs over the life of the vehicle are considered. Many buyers want a low purchase price more than they want a low operating cost, and for this reason high-efficiency cars will not automatically take over a large share of the market. Additional pressure in that direction will be required. Many suggestions have been made about which combinations of efficiency regulations, energy subsidies, tax incentives and penalties aimed at efficiency improvements, and public relations would be politically feasible and at the same time most effective in putting high-efficiency products into the hands of consumers and thereby reducing fossil-fuel use.

But, like the coal miners and mine owners who maintain that acid rain is not serious and climate heating is science fiction, automobile companies in America sometimes argue that light, efficient cars are more dangerous on the road than their larger, heavier products. The safety record of well-designed light cars does not support this contention, but the fact that this argument is put forward illustrates some of the political hurdles that stand in the way of rapid efficiency improvements.

Natural gas could play an anomalous role in an effort to reduce carbon dioxide emissions; that is, its use could increase. If, for example, countries decide that the easiest way to achieve the required reduction in carbon dioxide emissions is to tax emissions of carbon, natural gas would be in a favored position because it yields more energy for the same amount of carbon than coal or petroleum products do. But natural-gas deposits that are located far from urban areas may be of little use—the cost of pipelines to bring the gas to the consumer is too large—or burned as a waste product. Some regions with severe urban air-quality problems are experimenting with methanol as a replacement for gasoline in automobiles. Methanol can be synthesized from natural gas by processing plants near the deposits and then shipped as a liquid to the users, thus replacing gasoline and reducing net carbon dioxide emissions.

In moving toward increased gas use, manufacturers, processors, shippers, and government agencies will have to proceed cautiously in order to ensure that greater reliance on methane does not result in increased leakage to the air. Because the methane molecules is twenty or thirty times as effective in infrared trapping as the carbon dioxide molecule, leaks of methane to the air could cancel out the advantages from reduced carbon dioxide emissions. Methane also has secondary effects that contribute to atmospheric problems. It enhances reactions that create ozone, another infrared-trapping gas, in the lower atmosphere, and it is transformed in the stratosphere into water vapor, thereby hastening changes in the high atmosphere.

In this discussion of efficiency, we have focused so far on the industrial world. The United States, the Soviet Union, and Western Europe emit more than half of all the fossil-fuel CO_2 now being placed in the atmosphere. But experts studying the possibilities of fuel savings claim that opportunities for major improvements exist everywhere in the world, even in developing countries with low per capita fuel usage. These claims are sometimes disputed; intuitively it would seem that the rich countries must be the wasteful ones and that the search for efficiency improvements should start there. A committee of the United Nations Commission on the Environment, however, writing about energy use, supported the experts:

> These claims [of possible energy savings] are often rejected by developing countries, and the poor generally, as concerns of only the extravagant and well-to-do. Nothing could more grievously misrepresent the truth. It is the poorest who are most often condemned to use energy and other resources least efficiently and productively, and who can least afford to do so.
>
> The woman who cooks in an earthen pot over an open fire uses perhaps eight times more fuel than her affluent neighbor with a gas stove and aluminum pans. The poor who light their homes with a wick dipped in a jar of kerosene get one hundredth of the illumination of a 100-watt bulb and use just as much energy to do so.[53]

The committee points out ways in which developing countries can reduce energy consumption "without loss of welfare or output." The report adds, "But in a poor country the benefits thus gained will mean much more." A recent study at the Center for Energy and Environmental Studies at Princeton University, although it does not include fossil fuel, dramatically illustrates this point. Many Third World countries grow sugarcane and burn the residue as fuel at the sugar factories. In doing so, they produce about 20 kilowatt-hours of electricity per ton of residue, or about enough to run the factory. Were the residue to be gasified and burned in turbines similar to jet engines, 460 kilowatt-hours per ton could be produced and the factory could export electricity. Cane is widely grown, and this technology could produce up to one-fourth of the electricity now used in the seventy sugar-producing developing countries, allowing a 25 percent expansion of energy services there without new fossil-fuel exploitation.[54]

This example illustrates the issues to be confronted if needed Third World development is to take place without adding to atmospheric change. Development of these turbines occurred in the United States largely as a result of United States' lead in jet-engine technology, which in turn was fueled by the large investments made to keep this nation a leader in both military and civilian aircraft technology. The use of these turbines with natural-gas fuel is growing, and the effectiveness of gasified coal as fuel has already been demonstrated. At this stage, the additional effort needed to demonstrate the use of turbines with gasified biomass, such as sugarcane residue, is judged to be modest. And the costs are reasonable: the cost per kilowatt of capacity for the natural-gas turbines that have already been installed is about half that of new coal-fired plants, and the cost of the gasified coal plants will be competitive with those of current coal-fired plants. The biomass plants may be cheaper than the coal versions; they do not have to contend with the sulfur that plagues the design of gasified coal systems.

What is our best course of action? Develop the biomass

plants in the hope that generally poor developing countries will buy our technology? Share our current knowledge and designs with technologically underdeveloped countries, hoping they will carry forward the development and application of this efficient technology? Make such power plants part of our foreign aid? Perhaps a combination of strategies, each addressing the special problems and opportunities of the countries involved, will be required. But, at the very least, technical assistance and resource transfers from the industrial world will be needed if the developing countries are to join in the international movement toward more efficient energy use.

There is some stirring on the international scene. In Toronto in 1988 the Canadian government hosted a meeting of experts from around the globe to discuss "The Changing Atmosphere." The statement issued by this conference stressed the importance of reduced emissions in slowing the rate of climate change so that the world can adapt more smoothly to those changes that are already under way. The conference report recommended a 20 percent reduction in global carbon dioxide emissions by the year 2005. Studies focusing on this goal indicate that the technology that would make such a reduction possible is readily available and that the savings in resources would be large. Compared to the costs of constructing new power plants and paying for fuel in a "business as usual" scenario, meeting the Toronto goal in the OECD countries would produce annual savings of $35 billion in the industrial sector, $20 billion in transportation, and almost $50 billion in buildings by the year 2000.[55] The Toronto conferees had speculated that half the reduction goal could be met by efficiency improvements and half by substituting nonfossil energy sources for fossil fuels, but the early indications are that steady efficiency improvements of about 3 percent per year can achieve the same end without requiring a search for new energy sources.

But new energy sources will be needed—efficiency improvements can only carry us so far. Thus, the global strategy for CO_2 reductions also requires aggressive efforts to find inexpensive, acceptable, noncarbon alternative energy sources that will allow emissions reductions to continue after the economically desirable efficiency improvements have been fully exploited.

The candidates for new energy sources include a wide variety of existing technologies: solar photovoltaics, which, though falling rapidly in price, have not yet entered mass production and are therefore not yet competitive with coal-fired electricity; solar thermal, now making inroads into the energy market in clear desert locations; nuclear, which works, but which has yet to convince the public of its safety and finds its costs escalating as the problems of waste disposal and decommissioning become more apparent; and energy from temperature gradients in the earth or ocean, already in regular use in special places on land but not yet adequately demonstrated for more general application. Less likely candidates include nuclear fusion, whose practical demonstration always seems to be several decades in the future.

In this search for future energy sources, we can profit from past experience. We should not narrow the choices too soon, as we did with nuclear power, rather we should consider adopting a variety of approaches. We should think hard about the possible undesired side effects of each source and choose sources that produce the least amount of unrecyclable wastes. And these efforts should be matched by effective steps to stabilize the earth's population so that improvements in energy efficiency are not canceled out by a steady growth in our numbers.

FORESTS

One-fourth of the carbon dioxide entering the air today comes from biotic sources, mostly from deforestation. Slowing the conversion of forest to fields and limiting the harvesting of timber to the level at which the forest can regenerate itself could reduce this source of carbon dioxide and contribute substantially to an overall low-impact strategy for the future. Tropical forests, which contain most of the world's wood and living species, are the main concern.

In tropical countries, protecting these forests has proved politically troublesome. At the level of central government, exploitation of the forests for export markets may seem necessary, given the debt load of the country. At the level of corporations and entrepreneurs, the quick profits to be made from harvesting valuable trees are tempting. At the level of individuals and families, a forest may be viewed as the source of fuelwood and other necessities or it may represent potential farmland. Several nongovernmental groups are hard at work attempting to alert villagers to the long-term advantages of maintaining nearby forests and to instruct them in the principles of woodlot management and agroforestry. These measures can make a difference, but, as in the case of energy efficiency, improvements in local forestry practices can be nullified if populations continue to grow, and government actions can override local progress.

In some of the countries with large forest reserves, rapid deforestation is being stimulated by counterproductive government policies. Tax and credit incentives intended to create jobs, open up new land for development, or relieve urban crowding have had the effect of subsidizing wasteful logging practices, encouraging exports at subsidized prices (thus sacrificing possible national income), and converting forests into farms or ranches. Sadly, many of these policies have not been effective in solving the problems that led to their institution in the first place. Replacing these programs with carefully designed regulations, fees, and royalties, as well as providing reasonable monitoring and enforcement of the regulations, could make a large difference in the rate of deforestation. Misguided policies are not limited to tropical developing countries; the United States continues to sell logs abroad at prices that do not even cover the cost of getting them out of the forest and ready for shipment.[56]

POPULATION STABILIZATION

Each problem discussed here, and indeed almost every global problem facing us today, is driven in the first instance and exacerbated by growth in the world's population. Each year the nations of the earth collectively add more people to the total population than they did the year before; the only sign of progress toward stabilization is the fact that the world's population could have grown even faster but did not. In recent years the United States has added 1.7 million people to its population each year, Mexico 2 million, Bangladesh 2.9 million, Indonesia 3 million, Pakistan 3.1 million, Nigeria 3.2 million, and China 15.2 million. The highest percentage increase occurs in Kenya, which adds 4.1 percent but only 955,000 people each year; the greatest absolute growth takes place in India, which adds 16.3 million people a year for a 2 percent rate of growth.[57]

The issues of population growth and its effect on the atmosphere are intertwined with rates of consumption. A citizen of the United States uses ten to fifteen times as much energy as a citizen of India or China; therefore, reducing the United States' population growth by 1 million would lower CO_2 emissions as much as or more than reducing India's or China's growth by 10 million. Furthermore, if the United States chose to reduce its population growth by 1 million, its highly developed health-care system would make such a goal more realistic than an attempt by a large Third World country to reduce its population growth by 10 million. Yet as developing countries struggle to achieve a higher standard of living, continuing increases in their populations not only undermine their development goals but also guarantee higher energy consumption once those goals are attained. Thus, both rich and poor countries have reason to work toward stable or decreasing population numbers.

Lessons for a workable program to stabilize or reduce the world's population come from Denmark, Germany, Hungary, Italy, and the United Kingdom, which have reached stable populations, or from countries that have made notable progress in recent years such as Chile, China, Cuba, Singapore, South Korea, and Taiwan.[58] The components of a successful program vary with the level of prosperity of the society: the more prosperous a society, the more effective such straightforward programs as sex education and access to a variety of birth-control techniques will be in limiting fertility. In Third World countries, successful programs must be adapted to local circumstances and require more effort.[59]

The specific strategies for reducing fertility in Kenya, where women have an average of eight children in their lifetimes and say they want more, must differ from those in parts of Asia, where, it can be argued, a preference for sons contributes to keeping fertility above the rate required to produce zero growth. But everywhere in the Third World, reducing fertility, and hence population growth, requires effective economic development efforts, government approval, and energetic family-planning programs. The experience of the now industrialized world and the more recent history of developing countries show that as a country advances economically, fertility falls. But in some especially poor countries, rapid population growth is a bar to the very economic development that it is hoped will slow population growth.

This deadlock could be broken, at least in part, by the economic benefits of family planning. Family planning is an essential element of maternal and child health care: widely spaced children and their mothers are healthier and live longer than closely spaced children and their mothers. And better health acts as an incentive to limit family size: parents who develop confidence that their children will survive have fewer children. Fewer but healthier children will build stronger national economies; this reasoning should inspire even poor countries to work for slower population growth.

Experts know the basic features of a successful family-planning program. It should offer the full range of contraceptive techniques; it functions best in the context of a broader primary health-care program; and, above all, it must be appropriate to a country's cultural setting. In some countries the participation of trained, local women as proponents and teachers is essential. A symbolic commitment from the highest level of government to family planning, not only as a health measure but also as a means of reducing fertility and stabilizing population growth, is of critical importance to the program's acceptance and effectiveness. Also crucial to progress in family planning are advances in the economic, political, and educational status of women. Higher literacy rates; greater attendance in primary, secondary, and postsecondary schools; and broader opportunities for paid work outside the home—all expand women's options beyond early marriage and early, repeated childbearing. In addition to improving the lives of women, efforts like these would reduce fertility as well.

At this writing, the total annual investment from all sources in developing-country population programs is $3.2 billion. (Half a billion of this is contributed by foreign donors, a figure amounting to around 2 percent of total foreign aid.) Enhancing these programs and expanding them in countries with large populations and high growth rates, with the aim of substantially reducing population growth, will require that greater resources be allocated to these tasks.[60] Surveys show that 75 percent of married women in developing countries want to limit or space future births. It is estimated that an annual outlay of $10.5 billion, about three times the current expenditure, could make family-planning services available to these women. Most industrial countries can easily afford the needed programs in their own countries; elsewhere, increased foreign aid will be required. But the first step toward a stable population in the United States, as in some developing countries, is a commitment at the highest levels of government to that goal.

A global strategy to reduce the total human impact on the atmosphere thus reaches far beyond the usual concerns of atmospheric scientists or air-quality regulators. No such strategy has yet been fully described, but it would certainly include all the elements touched on in this book and more. It would involve stabilizing populations so as to make long-term solutions possible. It would involve prompt measures to improve efficiency in the generation and use of power, since these measures can be rapidly implemented and will bring other advantages to countries that do so. It would include development of future sources of energy that would enable us to reduce carbon dioxide emissions to below the rate at which the oceans can absorb them, without adding new problems to the earth and air. Reforestation programs and a reduced rate of deforestation could play an important role in the strategy. Greater care would need to be exercised in analyzing the impact of new synthetic gases before they are manufactured in large quantities and released into the atmosphere. Recycling such gases—indeed, recycling a much larger proportion of all the resources we use—will be an essential element. And, recognizing that we have already committed ourselves to some rate of atmospheric change, any strategy for the future must include measures designed to improve our ability to foresee where these changes are leading us and plans to counteract or adapt to a shifting climate.

Any reduction of the impacts is worthwhile, because it will

allow us a bit more time in which to adapt to the changes we are producing. But a more ambitious goal—reducing impacts to such an extent that significant portions of the evolved ecosystems can continue to function without human management—would require a large and rapid change. The beginning steps would be the same as those already described, but the goal of stabilizing the composition of the air would require major decreases in all emissions and leaving large tracts of forest, grassland, wetland, and tundra untouched and unmanaged.

This path that leads to decreasing the human impact on earth thus demands a constant vision of a nebulous long-term goal, major changes in actions and attitudes, and an unprecedented amount of international cooperation. The choice between our present course, with its dangers, and the difficult strategy needed to bring the human impact into equilibrium with the earth's oceans, atmosphere, and ecosystems is barely ours to make. Continued expansion, with its accompanying trust in technological solutions, is firmly established as our custom; any change would require major and continuing efforts on the part of world leaders in many fields and the development of a widely held, alternative definition of what it means to be human on earth. But the advantages of changing are great: the continuation of the evolved systems of the earth and adequate time for people to adapt constructively to inevitable changes.

There is one further advantage that a slower path would offer over our present one: it leaves our options open. If, some years or centuries hence, we are sure we can wisely manage the whole show, we can choose to do so. If, however, we proceed down the faster track, we will have no other choice but to continue on it. The natural systems will be gone and the job of constructing and managing a new world will continue to be ours for an indefinite time to come.

Notes

1. The Atmosphere and People

1. This brief discussion of the origin and evolution of the atmosphere draws heavily from J. C. G. Walker, *Evolution of the Atmosphere* (New York: Macmillan, 1977).

2. For a wide-ranging account of interactions between the air and life, see Stephen H. Schneider and Randi Londer, *The Coevolution of Climate and Life* (San Francisco: Sierra Club Books, 1984).

2. Acid Rain

3. The *National Geographic* photo-map, entitled "Portrait U.S.A.," was distributed with the July 1976 issue of the magazine.

4. For a recounting of the history of the Ducktown–Copper Hill smelters, as well as the comment about Queen Elizabeth's distaste for coal burning, see Robert E. Swain, "Smoke and fume investigations," *Industrial and Engineering Chemistry* 41 (1949): 2384–88.

5. The summary given here of the early history of acid-rain studies depends heavily on an extensive review of the topic: Ellis B. Cowling, "Acid precipitation in historical perspective," *Environmental Science and Technology* 16 (1982): 110A–23A. This article not only recounts the history but also describes the U.S. research and monitoring program designed to improve our knowledge of this subject.

6. A popular account of the deadly sulfate episodes in New York City is Roy Popkin, "Two 'killer smogs' the headlines missed," *EPA Journal* 12 (1986): 27–29.

7. The data on emissions of sulfur dioxide and nitrogen oxides are taken from a table in World Resources Institute and International Institute for Environment and Development, *World Resources 1987* (New York: Basic Books, 1987), 147. This table shows that sulfur oxide emissions in the United States increased sharply from 1940 to 1973, reaching a peak of 28.7 million metric tons per year. They then decreased to 21.4 mmt by 1984. Nitrogen oxide emissions, however, increased by 190 percent between 1940 and 1984 and even in the last decade, with controls on many auto exhausts, either continue at the same level or are increasing slowly.

8. An early, careful accounting of the human contribution to sulfur in the air and oceans, as compared to natural sources, is to be found in W. W. Kellogg, R. D. Cadle, E. R. Allen, A. L. Lazrus, and E. A. Martell, "The sulfur cycle," *Science* 175 (1972): 578–96. These writers conclude that human activities were contributing about half as much as nature to the total atmospheric burden of sulfur compounds, but that by the year 2000 such activities "will be contributing about as much." They point out that their estimate is quite uncertain, because so little was known about the various natural sources.

More recent studies, such as Meinrat O. Andreae and Hans Raemdonck, "Dimethyl sulfide in the surface ocean and the marine atmosphere: A global view," (*Science* 221 [1983]: 744–47), and M. V. Ivanov, "The global biogeochemical sulfur cycle (in *Some Perspectives of the Major Biogeochemical Cycles*, ed. G. E. Likens, SCOPE Report 17 [New York: Wiley & Sons, 1981]), show much progress in measurements of the ocean's sulfur chemistry and sulfur sources, but the overall numbers still hover around an approximate equality between natural and anthropogenic contributions to gaseous sulfur emissions to the atmosphere.

9. The measurements of lead in the Greenland ice are replotted from a graph in M. Murozumi, T. J. Chow, and C. Patterson, "Chemical concentrations of pollutant lead aerosols, terrestrial dusts and sea salts in Greenland and Antarctic snow strata," *Geochimica et Cosmochimica Acta*, 33 (1969): 1247–94. This report clearly shows a large increase in the amount of lead falling on Greenland. It also describes how difficult such measurements are. The authors comment that "polar ices are extremely pure, being nearly equal to the purest laboratory water." Large samples of ice had to be obtained in order to measure such small amounts of lead, while at the same time extreme precautions had to be observed to prevent contamination of the sample. A single hair from the head of a

workman who had driven a vehicle fueled with leaded gasoline or who had consumed soft drinks from cans sealed with solder would contain more pollutant lead than the hundreds of pounds of ice of a sample.

10. The sulfate measurements in Greenland are reported in M. M. Herron, "Impurity sources of F^-, Cl^-, NO_3^- and SO_4^{2-} in Greenland and Antarctic precipitation," *Journal of Geophysical Research* 87, no. C4 (1982): 3052–60.

11. A recent summary of what is known about the impacts of acid rain on terrestrial ecosystems is to be found in D. W. Schindler, "Effects of acid rain on freshwater ecosystems," *Science* 239 (1988): 149–56, and James J. MacKenzie and Mohamed T. El-Ashry, *Ill Winds: Airborne Pollution's Toll on Trees and Crops* (Washington, D.C.: World Resources Institute, 1988).

3. Stratospheric Ozone

12. The relationship of ozone to life brings up a scientific question that has yet to be satisfactorily resolved: Which came first, ozone or life? Ozone is a form of oxygen—three atoms in a molecule instead of the usual two—and oxygen was put into the air by plants in the process of photosynthesis. But without ozone to protect them from strong ultraviolet light, how could plants ever become common enough to supply the required oxygen? And without plants, how could the high atmosphere ever acquire enough oxygen to make ozone?

A similar question arises from attempts to imagine the changes that had to happen when photosynthesis first got going in a big way. The waste product of photosynthesis is oxygen, but until that time living things did not have to deal with such a chemical. Oxygen is now normal and necessary, but it is in fact a highly corrosive substance that rusts iron, bleaches colors, and rots biological material. Early plant life would not have developed the techniques that modern forms of life have of canceling the harmful effects of oxygen; in other words, oxygen would have been a toxic gas at that time. So how was the transition managed?

The answer to both of these questions probably lies in the protection provided by the oceans, the very long times involved in the transitions, the great variety of life that always seems to occur, and, in that variety, the existence of some life forms more resistant to ultraviolet light or oxygen corrosion than others.

13. The worries about the depletion of the ozone layer did not start

with the realization that the chlorine of CFCs could catalyze ozone destruction. A few years earlier, scientists had argued that nitrogen oxides in the exhaust output of high-flying aircraft—in particular the various supersonic craft, or SSTs—could place chemicals at the right altitude to do damage. This episode was followed by studies aimed at determining whether the chlorine in the fuel used to launch the U.S. Space Shuttle, some of which would be deposited in the ozone layer, would do unacceptable damage. These concerns were soon overtaken by those about CFCs, which are produced in much larger quantities than either SST exhaust or shuttle fuel. A lively accounting of these three episodes is contained in Lydia Dotto and Harold Schiff, *The Ozone War* (Garden City, N.Y.: Doubleday, 1978). A more recent account is to be found in John Gribbin, *The Hole in the Sky* (New York: Bantam Books, 1988).

14. The graphs showing the concentrations of CFC-11 and CFC-12 were prepared from information in World Meteorological Organization, *Atmospheric Ozone 1985*, Report 16 of the Global Ozone Research and Monitoring Project WMO, (Geneva: WMO, 1985), 59–63, and updated through 1987 with data kindly supplied by Ron Prinn at MIT.

15. The total ozone curves for Arosa, Switzerland, were reported by one of the pioneer research workers in ozone studies in Hans U. Dütsch, "Vertical ozone distribution over Arosa," Technical Report, National Center for Atmospheric Research, Boulder, Colorado, 1964.

16. A summary of the biological effects of ozone depletion is to be found in James G. Titus, ed., *Effects of Changes in Stratospheric Ozone and Global Climate*, vol. 2 (United Nations Environmental Programme and the U.S. Environmental Protection Agency, 1986).

17. The estimates of the effect of the treaty to limit CFC and halon production and consumption come from a staff paper by the Oceans and Environment Program of the U.S. Office of Technology Assessment, entitled "An Analysis of the Montreal Protocol on Substances that Deplete the Ozone Layer," and dated February 1, 1988.

4. Climate Heating

18. A calibration curve, similar to the one in figure 7, is displayed in W. Libby, *Radiocarbon Dating*, 2d ed. (Chicago: University of Chicago Press, 1955). The curve is calculated from the half-life of ^{14}C as measured in the laboratory. Libby shows that the curve agrees reasonably well with a few well-dated archaeological specimens, such as a beam from the

tomb of Vizier Hemaka in Egypt and the wrappings of the Dead Sea Scrolls.

19. The discovery of the deficiency of ^{14}C in modern wood is described in H. E. Suess, "Radiocarbon concentration in modern wood," *Science* 122 (1954): 415–17. Figure 8 is simply a schematic I prepared to illustrate what Suess suspected was the case. In the years following 1954, Suess and others showed that the actual calibration curve was quite complicated, as shown in figure N1. Some authors began to refer to these deviations from a simple form as "Suess wiggles." Today this sort of calibration curve is determined by measuring the ^{14}C in wood from tree rings of known age and comparing the results with age calculated from the simple calibration curve. But the fact that the "wiggles" are almost half as large as the modern downturn makes the story of the discovery of the Suess effect all the more remarkable. If he had happened on tree rings from 1600 to 1700, for example, he might have had more difficulty in reaching his conclusions. The data used in figure N1 are from *Radiocarbon* 28, no. 2B (1986), and the establishment of the wiggles as a real effect

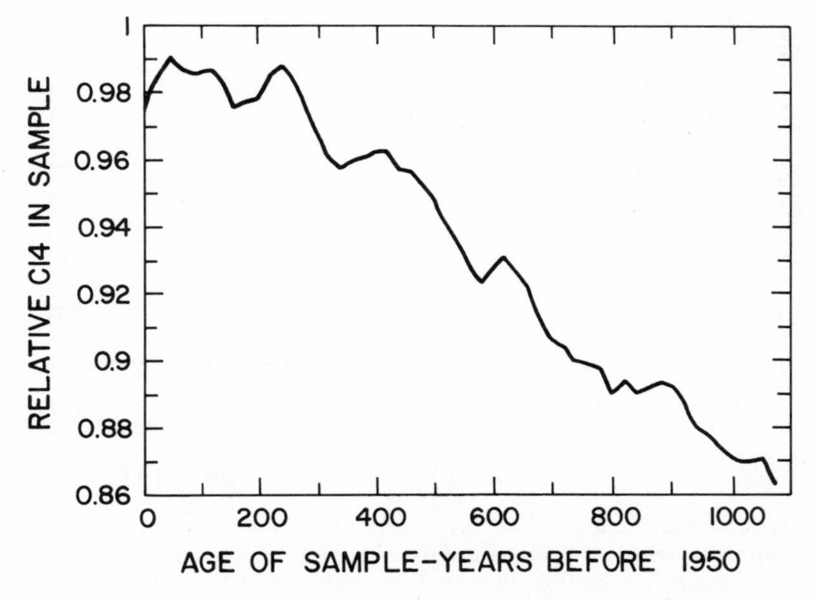

FIGURE N1. The calibration curve for radiocarbon dating measured from the ^{14}C content of well-dated tree rings grown before 1950.

is described in H. E. Suess, "Radiocarbon geophysics," *Endeavor*, n.s. 4 (1980): 113–17.

20. The famous quotation about the "large scale geophysical experiment" comes from Roger Revelle and Hans E. Suess, "The question of increase of atmospheric CO_2," *Tellus* 9 (1957): 18–27.

21. It has been more than thirty years since the final growth of the samples that Suess used for his discovery, and during that time people have burned more fossil fuel than in all the preceding centuries. Has the Suess effect increased rapidly? In fact the effect has, perhaps temporarily, quite vanished under the deluge of extra ^{14}C placed in the atmosphere by the above-ground testing of nuclear weapons during the 1950s and 1960s as shown in figure N2 (data from various articles in the journal *Radiocarbon*). Instead of fluctuating a percent or two from the

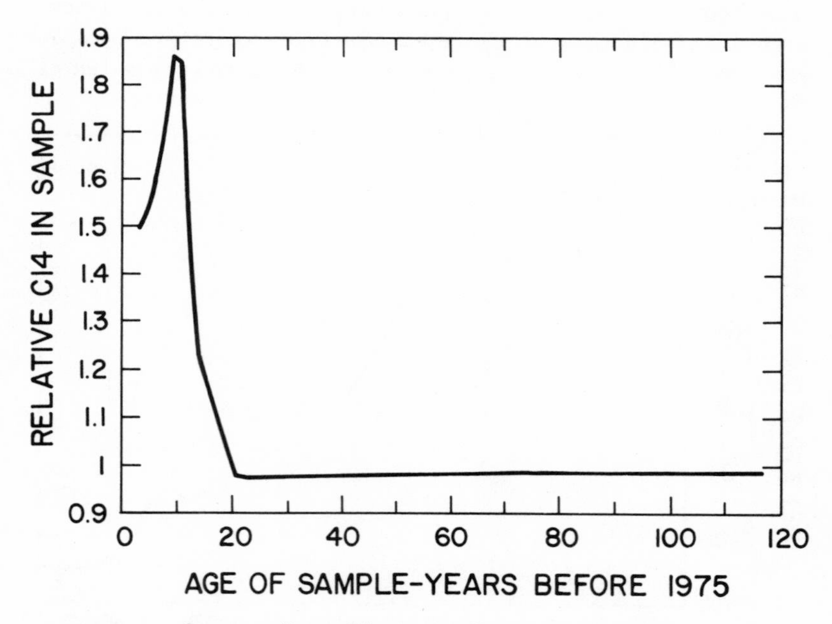

FIGURE N2. Same as figure N1, except for material grown up to approximately 1975. The part of the curve for 120 to 20 years before 1975 shows the same slight downward trend of the Suess Effect as in figure N1; the most recent years show the large amount of ^{14}C put into the air by tests in the atmosphere of nuclear bombs.

simple calibration curve, as is the case with the Suess effect, these measurements indicate changes of almost 100 percent. It seems that ^{14}C is becoming less and less an indicator of age and more and more a record of the industrial and military prowess of the human race.

22. The measurements of carbon dioxide concentrations in the atmosphere shown in figure 9 were made by C. David Keeling of the Scripps Institution of Oceanography, La Jolla, California, and the data are archived at the Carbon Dioxide Information Center, Oak Ridge National Laboratory, Oak Ridge, Tennessee.

23. The ice-core measurements of carbon dioxide are from data displayed in H. Friedli et al., "Ice core record of the $^{13}C/^{12}C$ ratio of atmospheric CO_2 in the past two centuries," *Nature* 324, (1986): 237–38.

24. The idea that the burning of fossil fuel could heat the climate is usually attributed to Svante Arrhenius, a Swedish scientist who was active around the turn of the century. But in his book *Worlds in the Making*, trans. H. Borns (New York and London: Harper & Brothers, 1908), Arrhenius gives credit to "the great French physicist Fourier" for the idea that atmospheres could trap heat. Arrhenius recognized that the amount of coal use, even at the end of the nineteenth century, was large enough to have an appreciable effect on the atmospheric concentration of carbon dioxide. He went on to estimate how much heating would result from a doubling of carbon dioxide in the atmosphere. His estimates are well within the range of results from modern model calculations—1.5° to 5.5° C.

25. The direct measurements of methane in the air, replotted in figure 11, are from R. A. Rasmussen and M. A. K. Khalil, "Atmospheric methane in the recent and ancient atmospheres: Concentrations, trends, and interhemispheric gradient," *Journal of Geophysical Research* 89, D7 (1984): 11599–11605; and D. R. Blake and F. S. Rowland, "World-wide increase in tropospheric methane, 1978–1983," *Journal of Atmospheric Chemistry* 4, (1986): 43–62.

The ice-core measurements of methane used in figure 11 are reported in B. Stauffer, G. Fischer, A. Neftel, and H. Oeschger, "Increase of atmospheric methane recorded in antarctic ice," *Science* 229 (1985): 1386–87.

26. A description of how climate models are constructed is given in W. M. Washington and C. L. Parkinson, *An Introduction to Three-dimensional Climate Modeling* (Mill Valley, Calif.: University Science Books, 1986).

27. I make several references to water vapor as a powerful infrared-trapping gas, but no comment on how its concentration in the atmosphere might change. Water does not stay in the air very long; when the relative humidity gets above a certain point it usually rains. Most models therefore attempt to simulate the amount of water vapor in the air as controlled by the temperature of the air, and hence by the amount of long-lived infrared-trapping gases. Thus, water vapor shows up as a positive feedback in the climate calculations: if it gets hotter, the air can hold more water vapor, which makes it hotter still. This feedback process was recognized quite early—Arrhenius included it in his calculation.

28. The topic of climate heating is periodically reviewed by groups of scientists who meet to study progress since the last review and write a report on their findings. The latest international report, and in many ways the most comprehensive, is from the 1985 Villach Meeting, named for the town in Austria in which the meeting was held. See Bert Bolin, Bo R. Döös, Jill Jager, and Richard A. Warwick, *The Greenhouse Effect, Climatic Change, and Ecosystems*, SCOPE Report 29 (Chichester: Wiley & Sons, 1986).

29. The review of the level of agreement among models of climate heating and the range of possible temperature increases if carbon dioxide doubles is contained in a chapter in the 1985 Villach report; see R. E. Dickinson, "How will climate change?" chap. 5 in *The Greenhouse Effect, Climate Change, and Ecosystems*, ed. B. Bolin et al. (Chichester: Wiley & Sons, 1986).

30. In the United States, climate change is reviewed periodically by committees of the National Academy of Sciences. The report discussed here is *Changing Climate: Report of the Carbon Dioxide Assessment Committee* (Washington, D.C.: National Academy Press, 1983).

31. Jill Jaeger, *Developing Policies for Responding to Climatic Change*, WMO/TD no. 225 (Geneva and Nairobi: World Meteorological Organization and United Nations Environment Program, 1987).

32. The temperature measurements at Saint Louis, Denver, Columbus, and Palma are from the data archives of the National Center for Atmospheric Research, Boulder, Colorado.

33. The Palma measurements represent a smaller geographic area than do the Saint Louis observations. Locations in Italy and Greece do not show the large drop in temperature in the 1930s.

34. The surface temperatures of the Northern and Southern Hemi-

spheres since 1860 are taken from P. D. Jones, T. M. L. Wigley, and P. B. Wright, "Global temperature variation between 1861 and 1984," *Nature* 322 (1986): 430–32.

35. The data used to plot the temperatures during the last one thousand years were taken from H. H. Lamb, *Climate: Present, Past, and Future,* vol. 2 (London: Metheun, 1977), 564, and from V. C. La Marche, Jr., "Tree-ring evidence of past climatic variability," *Nature* 276 (1978): 334–38.

36. A summary of the evidence concerning the rise and fall of the Norse settlements in Greenland is given in T. H. McGovern, "Economics of extinction in Norse Greenland," in *Climate and History,* ed. T. M. L. Wigley (Cambridge: Cambridge University Press, 1981).

37. The skeptical historian was Emmanuel Le Roy Ladurie, *Times of Feast, Times of Famine: A History of Climate Since the Year 1000* (Garden City, N.Y.: Doubleday, 1971).

38. The seventeen-thousand-year record was taken from data in C. Lorius, L. Merlivat, J. Jouzel, and M. Pourchet, "A 30,000-year isotope climatic record from Antarctic ice," *Nature* 280 (1979): 644–48, and S. J. Johnson, W. Dansgaard, H. B. Clausen, and C. C. Langway, "Oxygen isotope profiles through the Antarctic and Greenland ice sheets," *Nature* 235 (1972): 429–35.

39. The migrations of forests as the last ice sheet retreated are described in J. C. Bernabo and T. Webb III, "Changing patterns in the Holocene pollen record of northeastern North America: A mapped summary," *Quarternary Research* 8 (1977): 64–96.

40. A review of estimates of possible loss of coastal wetlands in the coming decades is found in J. G. Titus, "The causes and effects of sea level rise," in *Effects of Changes in Stratospheric Ozone and Global Climate,* vol. 1, ed. J. G. Titus (U.S. Environmental Protection Agency, 1986), 219–48.

5. But Is It True?

41. Figure 20 was supplied by Dr. Warren Washington of the National Center for Atmospheric Research. The model simulations used to produce the left-hand portion of the chart are from E. J. Pitcher, R. C. Malone, V. Ramanathan, M. L. Blackmon, K. Puri, and W. Bourke, "January and July simulations with a spectral general circulation model,"

Journal of Atmospheric Science 40 (1983): 580–604. The model used for these calculations was the NCAR Community Climate Model, Version O. The data summaries used to construct the right-hand side of figure 20 are taken from C. Schutz and W. L. Gates, "Global climatic data for surface, 800 mb, 400 mb: January," Report R-915-ARPA (Santa Monica, Calif.: Rand Corp., 1971), and C. Schutz and W. L. Gates, "Global climatic data for surface, 800 mb, 400 mb: July," Report R-1029-ARPA (Santa Monica, Calif.: Rand Corp, 1972).

42. The comparisons of actual climates since the retreat of the last ice age, as reconstructed from proxy data, with model climates, is described in COHMAP Members, "Climatic changes of the last 18,000 years: Observations and model simulations," *Science* 241 (1988): 1043–52.

Sometimes the tests of models have unexpected outcomes. It is well known that some sixty million years ago, when dinosaurs roamed and coal was being formed, the climate was much warmer than it is now. It has also been deduced that continents drift over the surface of the earth, and their positions during this Cretaceous period can be plotted. Modelers regarded this situation as a good one for asking the question, Would the positions of the continents as reconstructed for the Cretaceous make the climate as warm as fossil evidence seems to indicate that it was? This question was based on the knowledge that the oceans are an important part of the climate system, and that with a different continental configuration the flow of ocean currents would be completely different. It also relied on the discovery in central Asia of a Cretaceous fossil of a species that cannot tolerate freezing weather, which suggested that the climate was a great deal warmer then that now.

An advanced climate model was arranged with the continents in the new location and the climate computed; see E. J. Barron, "Climate models: Application for the pre-Pleistocene," in *Paleoclimate Analysis and Modeling*, ed. A. D. Hecht (New York: Wiley and Sons, 1985). The answer was that the climate was warmer, but frost still occurred in central Asia. Several ideas were put forward to explain this difference. Perhaps the atmosphere contained much more carbon dioxide then, and the added infrared trapping produced the extra heat; see E. J. Barron and W. W. Washington, "Warm Cretaceous climates: High CO_2 as a plausible mechanism," in *The Carbon Cycle and Atmospheric CO_2: Natural Variations Archean to Present*, ed. E. T. Sundquist and W. S. Broecker (Washington, D.C.: American Geophysical Union, 1985).

Some scientists suggested that the Asian fossil had been misinterpreted; perhaps the species of the fossil had been correctly identified, but it had changed enough in the long period of time since the Cretaceous that it could no longer tolerate freezing as it once did. This puzzle has not yet been solved, but it is interesting to note that all the suggested reasons for the lack of agreement between the model calculation and the fossil evidence assume that the model calculation is reliable and that the cause for the disparity must lie elsewhere.

43. A discussion of each of the major heat-trapping gases, their sources and sinks, and their rate of change in the atmosphere is to be found in World Meteorological Organization, *Atmospheric Ozone 1985*, Global Ozone Research and Monitoring Project Report no. 16 (Geneva, 1985).

44. For the Gaia hypothesis, see J. E. Lovelock and L. Margulis, "Homeostatic tendencies of the earth's atmosphere," *Origins of Life* 5 (1974): 93–103.

6. Just One Problem

45. The quoted editorial was published in the Boston Herald on July 26, 1937, and reprinted in *Bulletin of the American Meteorological Society* 18 (1937): 374–75.

46. The statement by Senator Watkins of Utah to David Brower of the Sierra Club appears in the report of the hearings on S.1555, the Colorado River Storage Project, before the Subcommittee on Irrigation and Reclamation of the Committee on Interior and Insular Affairs, 83d Congress, 2d sess., June 28–July 3, 1954, p. 520.

47. The quote about controlling the climate is from Jonathan Weiner, *Planet Earth* (New York: Bantam Books, 1986).

48. The failure of flood-control projects to achieve a reduction in flood damage is discussed in Gilbert F. White, *Human Adjustment to Floods*, University of Chicago Department of Geography Research Papers no. 29 (Chicago, 1945), and Gilbert F. White, Wesley C. Calef, James W. Hudson, Harold M. Mayer, John R. Schaeffer, and Donald J. Volk, *Changes in Urban Occupancy of Flood Plains in the United States*, University of Chicago Department of Geography Research Papers no. 57 (Chicago 1958).

49. A study of the impact of green revolution efforts in part of Mex-

ico is reported in Miguel Baraona, Guy Duval, Rolando Garcia, Susan Sanz, and Fernando Tudela, *Biospheric Changes and Food Systems* (Mexico, D.F.: Centro de Investigacion y de Estudios Avanzados de IPN, 1986).

50. I expect that the idea that the tragedy of air pollution is that it won't kill us has been around for some time, but I have been unable to trace its source.

7. The Other Path

51. The case for rapid improvements in energy-use efficiency is made in a number of publications, including John H. Gibbons and W. U. Chandler, *Energy, The Conservation Revolution* (New York: Plenum Press, 1981); William U. Chandler, *Energy Productivity: Key to Environmental Protection and Economic Progress,* Worldwatch Paper 63, (Washington, D.C.: Worldwatch Institute, 1985); John O. Blackburn, *The Renewable Energy Alternative* (Durham N.C.: Duke University Press, 1987); and Jose Goldemberg, T. B. Johansson, A. K. N. Reddy, and R. H. Williams, *Energy for a Sustainable World* and *Energy for Development* (Washington, D.C.: World Resources Institute, 1987).

These, and other studies pertaining to the situation in the United States, are summarized and converted to policy recommendations in W. U. Chandler, H. S. Geller, and M. R. Ledbetter, *Energy Efficiency: A New Agenda* (Washington, D.C.: American Council for an Energy-Efficient Economy, 1988).

52. For a discussion of the employment impact of moves toward energy efficiency, see U.S. Department of Energy, *Creating Jobs Through Energy Policy*, DOE/PE-0013, 1979.

53. The discussion of whether energy savings are possible only in rich societies is to be found in the report of the Panel on Energy of the United Nations' World Commission on Environment and Development, *Energy 2000: A Global Strategy for Sustainable Development* (London and New Jersey: Zed Books, 1987).

54. The example of the high-efficiency use of sugarcane residue is found in E. D. Larson and R. H. Williams, "Biomass-gasifier steam-injected gas turbine cogeneration," *Journal of Engineering for Gas Turbines and Power*, in press.

55. My discussion of energy efficiency relies heavily on the Goldemberg et al. report (listed above) for examples, and on discussions with

William Chandler concerning the possibilities for meeting the Toronto goal of a 20 percent reduction of CO_2 emissions by 2005. The example of efficient electrical generation using gasified sugarcane stalks and turbines comes from testimony by Robert H. Williams before the Subcommittee on Foreign Relations of the House Appropriations Committee, February 21, 1989.

56. The relationship of government policies to forest destruction is discussed in Robert Repetto, *The Forest for the Trees? Government Policies and the Misuse of Forest Resources* (Washington, D.C.: World Resources Institute, 1988).

57. Population increase numbers are taken from *1988 World Population Data Sheet,* Population Reference Bureau, Washington, D.C.; C. McEvedy and R. Jones, *Atlas of World Population* (New York: Penguin Books, 1978); and World Resources Institute and International Institute for Environment and Development, *World Resources 1988–89* (New York: Basic Books, 1988).

Because the United Nations and other organizations publish projections of global population showing a leveling sometime in the next century, and because we hear about decreasing global birthrates, people get the impression that the "population explosion" is diminishing. In fact, the absolute global population growth rate—the number of people added to the earth each year—is still increasing. During the period from 1965 to 1970, the world population increased by 56 million people per year; from 1975 to 1980, by about 75 million a year; and by 1987, by 85 million a year. Statisticians may mislead when they express population growth as a percent and note that the percentage has decreased since about 1967. What they fail to emphasize is that the percentage is applied to a base that is increasing faster than the percentage is decreasing; therefore, the number of people added each year continues to climb.

The longer-term projections by the United Nations and others that show a stabilization of the world population are based on the hope that the percentage increase can decline more rapidly than the population increases.

58. See note 57 above.

59. The lessons learned concerning family planning and population stabilization are described in Judith Jacobsen, *Promoting Population Stabilization: Incentives for Small Families,* Worldwatch Paper 54, (Washington, D.C.: Worldwatch Institute, 1983); Mead Cain *Women's Status and Fertility*

in Developing Countries, World Bank Staff Working Paper no. 682 (Washington, D.C.: World Bank, 1984); *World Development Report 1984* (Washington, D.C.: World Bank); Lester Brown, et al., *State of the World* (Washington, D.C.: Worldwatch Institute, 1985); and Lester Brown, et al., *State of the World* (Washington, D.C.: Worldwatch Institute, 1988).

60. The estimates of funds now spent on family planning in developing countries and the funds required to provide universal access to such services are taken from an unpublished memorandum prepared by The Population Crises Committee (Washington, D.C.: 1989).

Index